SpringerBriefs in Ethics

Springer Briefs in Ethics envisions a series of short publications in areas such as business ethics, bioethics, science and engineering ethics, food and agricultural ethics, environmental ethics, human rights and the like. The intention is to present concise summaries of cutting-edge research and practical applications across a wide spectrum.

Springer Briefs in Ethics are seen as complementing monographs and journal articles with compact volumes of 50 to 125 pages, covering a wide range of content from professional to academic. Typical topics might include:

- Timely reports on state-of-the art analytical techniques
- A bridge between new research results, as published in journal articles, and a contextual literature review
- A snapshot of a hot or emerging topic
- In-depth case studies or clinical examples
- Presentations of core concepts that students must understand in order to make independent contributions

Philip Brey

The Metaverse: A Critical Assessment

Philip Brey
University of Twente
Enschede, Overijssel, The Netherlands

ISSN 2211-8101　　　　　　ISSN 2211-811X　(electronic)
SpringerBriefs in Ethics
ISBN 978-3-031-93470-4　　　ISBN 978-3-031-93471-1　(eBook)
https://doi.org/10.1007/978-3-031-93471-1

© The Editor(s) (if applicable) and The Author(s), under exclusive license to Springer Nature Switzerland AG 2025

This work is subject to copyright. All rights are solely and exclusively licensed by the Publisher, whether the whole or part of the material is concerned, specifically the rights of translation, reprinting, reuse of illustrations, recitation, broadcasting, reproduction on microfilms or in any other physical way, and transmission or information storage and retrieval, electronic adaptation, computer software, or by similar or dissimilar methodology now known or hereafter developed.
The use of general descriptive names, registered names, trademarks, service marks, etc. in this publication does not imply, even in the absence of a specific statement, that such names are exempt from the relevant protective laws and regulations and therefore free for general use.
The publisher, the authors and the editors are safe to assume that the advice and information in this book are believed to be true and accurate at the date of publication. Neither the publisher nor the authors or the editors give a warranty, expressed or implied, with respect to the material contained herein or for any errors or omissions that may have been made. The publisher remains neutral with regard to jurisdictional claims in published maps and institutional affiliations.

This Springer imprint is published by the registered company Springer Nature Switzerland AG
The registered company address is: Gewerbestrasse 11, 6330 Cham, Switzerland

If disposing of this product, please recycle the paper.

Acknowledgments

This book would not have been possible without the support and inspiration of many individuals and institutions. I am grateful for the opportunity to present themes from this book during invited talks at the Computer Ethics Philosophical Enquiry conference at the Illinois Institute of Technology, Chicago; the Symposium Virtual and Augmented Realities at the Collège de France, Paris; the Information Ethics Roundtable on Ethics and Epistemology of Virtual Reality at Northeastern University, Boston; the Future of Digital Well-Being conference at the Royal Dutch Academy of Sciences in Amsterdam; and various other workshops and colloquia.

I have also benefited greatly from participation in the IEEE P7016 Standard for Ethically Aligned Design and Operation of Metaverse Systems and IEEE P2048 Standard for Metaverse: Terminology, Definitions, and Taxonomy working groups, and I am appreciative of the collegial exchanges.

I am also deeply grateful for the enriching exchanges and thoughtful feedback generously offered by peers and stakeholders around the world, including Lauri Kanerva and Nanayaa Appenteng (Meta), Hannah Claus, Pengyuan Zhou, Adam Henschke, Leonardo Werner, Wijnand IJsselsteijn, Matthew Dennis, Tilo Hartmann, Karen Phillips, and Teodora Spirova (UT VR lab).

Special thanks for excellent help and support at various stages of the manuscript from Samuela Marchiori, Tynke Schepers, Bianca Sutcliffe, Gamze Kaya, and Donagh Ó Brúdair.

The research presented in this book was partially supported by the research program *Ethics of Socially Disruptive Technologies*, funded through the Gravitation program of the Dutch Ministry of Education, Culture and Science, and the Netherlands Organization for Scientific Research (NWO grant no. 024.004.031).

Contents

1	**Introduction**		1
	References		4
2	**What Is the Metaverse?**		5
	2.1	A Short History of the Metaverse, from *Snow Crash* to the Present Day	6
		2.1.1 Neal Stephenson's Metaverse	7
		2.1.2 The Metaverse in Fiction	8
		2.1.3 Gaming, the Internet and Shared Virtual Spaces	8
		2.1.4 The Metaverse as a Developer Concept	9
		2.1.5 The Rise of Metaverse Technologies	10
		2.1.6 Mark Zuckerberg's Announcement and the Metaverse Initiative Boom	10
		2.1.7 Cooling-off and Realignment	13
	2.2	A Framework for Understanding the Metaverse	14
		2.2.1 Virtual Worlds: Unity, Interoperability, and Persistence	14
		2.2.2 Extension to Augmented Reality	16
		2.2.3 Immersiveness	17
		2.2.4 Multifunctionality	18
		2.2.5 Openness and Decentralization	18
		2.2.6 Relation to the Internet and to the Physical World	19
		2.2.7 Degree of Social, Economic, and Political Organization	20
	2.3	Key Metaverse Technologies	20
		2.3.1 3D Modeling and Animation	21
		2.3.2 Virtual Reality	21
		2.3.3 Augmented and Mixed Reality	23
		2.3.4 Artificial Intelligence	25
		2.3.5 Blockchain	26
		2.3.6 Internet of Things	27
		2.3.7 Infrastructure Technologies	28
		2.3.8 Conclusion	28

	2.4	So What Is the Metaverse?.	28
	References.		30
3	**Will There Be a Metaverse?**		**33**
	3.1	Introduction	34
	3.2	Why the Metaverse? Potential Benefits for Users and Society	35
		3.2.1 General Benefits for Users and Society	36
		3.2.2 Benefits within Specific Application Domains	38
	3.3	The Business Case for the Metaverse and Potential Barriers	41
		3.3.1 Potential Barriers	42
	3.4	Four Trends Toward the Metaverse	45
		3.4.1 The Metaverse and Web 3.0.	46
		3.4.2 The Metaverse and Natural Interaction	47
		3.4.3 The Metaverse and Online Worldbuilding.	48
		3.4.4 The Metaverse and Geographical Disembedding	52
		3.4.5 Other Trends and Conclusion	54
	3.5	Conclusion.	54
	References.		55
4	**Will the Metaverse Respect Our Rights?**		**59**
	4.1	Introduction	60
	4.2	Security in the Metaverse.	60
		4.2.1 Cybersecurity Threats in the Metaverse.	61
		4.2.2 Threats to Virtual Security in the Metaverse	62
		4.2.3 Safeguarding Security in the Metaverse	64
	4.3	Freedom in the Metaverse	65
		4.3.1 Freedom of Expression	65
		4.3.2 Bodily Integrity, Freedom of Movement, and Freedom of Assembly.	67
		4.3.3 Deception and Manipulation	68
	4.4	Privacy in the Metaverse	68
		4.4.1 New Privacy Risks in the Metaverse	69
		4.4.2 User Profiling, Tracking and Personalization in the Metaverse.	70
		4.4.3 Surveillance Capitalism in the Metaverse	71
		4.4.4 Privacy Protection in the Metaverse.	73
	4.5	Equality, Fairness, and Inclusion in the Metaverse	73
		4.5.1 Access to the Metaverse.	74
		4.5.2 Biases in the Metaverse	75
		4.5.3 User to User Discrimination	76
	4.6	Property Rights in the Metaverse.	77
		4.6.1 Protecting Property Rights in the Metaverse	78
		4.6.2 Ownership Models and the Common Good.	80
		4.6.3 Cryptocurrencies and NFTs.	81
	4.7	Conclusion.	82
	References.		83

5	**Will the Metaverse Benefit Humans and Society?**		85
	5.1	Introduction	86
	5.2	Well-being in the Metaverse	86
		5.2.1 Key Aspects of Well-Being and the Metaverse	87
		5.2.2 The Metaverse and Mental Health	88
		5.2.3 The Metaverse and Social Well-Being	91
	5.3	The Metaverse, Politics, and Civil Society	93
		5.3.1 The Social Quality of the Metaverse	93
		5.3.2 Democracy in the Metaverse	95
		5.3.3 The Metaverse and the Good of Society	96
	5.4	The Metaverse and Environmental Sustainability	97
		5.4.1 Negative Environmental Impacts	98
		5.4.2 Greening the Metaverse	99
		5.4.3 Environmental Savings Through Virtual Replacements	99
		5.4.4 Environmental Savings Through Environmental Simulations	102
	5.5	Conclusion	103
	References		104
6	**How Can the Metaverse be Developed Responsibly?**		109
	6.1	Introduction	110
	6.2	Metaverse Ethics and Multistakeholder Governance	110
		6.2.1 Principles of Responsible Governance	111
		6.2.2 Corporate-Driven Responsible Governance	111
		6.2.3 The Role of Other Stakeholders	112
	6.3	Responsible Development and Ethics by Design	115
		6.3.1 Ethics by Design	115
		6.3.2 Applying Ethics by Design to the Metaverse	117
		6.3.3 The Broader Ecosystem	118
	6.4	Responsible Platform Operation	118
		6.4.1 Ethics in Deployment and Operation	118
		6.4.2 Applying EDO to Metaverse Platforms	120
		6.4.3 The Role of Other Actors	121
	6.5	Conclusion	122
	References		122
Annex: A Moral Framework for Metaverse Ethics			123
Index			131

Chapter 1
Introduction

Abstract This brief introduction to *The Metaverse: A Critical Assessment* sets the stage for an in-depth examination of the metaverse—its definition, potential applications, enabling technologies, societal implications, and ethical concerns. It outlines the current state of metaverse development and public perception, offering a preliminary definition and previewing the book's structure. The book's aim is to inform stakeholders about the metaverse's opportunities and risks, providing practical guidance for its responsible development, use, and governance as a potentially transformative digital platform.

Keywords Metaverse · Immersive technology · Virtual reality · Governance · Ethics

The concept of the metaverse entered public discourse in the early 2020s. With Meta leading the charge, major tech companies promoted the metaverse as the next evolution of the internet—a transformation from symbolic, screen-based interactions to immersive, embodied experiences in which presence, movement, and interaction are digitally simulated and viscerally perceived. The metaverse was said to herald a new computing era, driven by the rise of immersive virtual and augmented reality systems designed to supplement or replace traditional screen- and keyboard-based interfaces. The internet was imagined to evolve into a three-dimensional interactive universe—the metaverse—where users could work, socialize, create, and play in dynamic virtual spaces beyond the boundaries of conventional browsing.

In the metaverse, it was suggested, a remote worker could collaborate in immersive 3D workspaces, making teamwork more engaging and interactive than traditional video calls. A medical student might practice surgical procedures in realistic virtual simulations before ever entering an operating room. A person with limited mobility could climb virtual mountains, or dance at concerts through their avatar. An aspiring entrepreneur might launch a virtual storefront, reaching global customers without the overhead of physical infrastructure. And a retired person could stay socially connected by joining virtual clubs, games, and events with friends and family around the world.

The metaverse presented a compelling vision of the future. However, tech companies did not deliver on it in the short term, as no fully realized metaverse emerged. Enthusiasm for the metaverse was soon overshadowed by the rise of a more immediately impactful technology: generative artificial intelligence, which, unlike the metaverse, was rapidly demonstrating real-world applications and delivering tangible results. In the media, the metaverse came to be seen either as a failed experiment or a distant, long-term project. Yet many in the tech community continued to believe in both the technological feasibility and economic viability of the metaverse, and significant investments have persisted in its enabling technologies, such as virtual and augmented reality, 3D modeling, blockchain, and artificial intelligence (AI). There remains a strong possibility that the metaverse—or a technological framework resembling it—will emerge in the coming years or decades, ushering in a transformative shift in computing and digital interaction.

This book offers a critical assessment of the metaverse—examining what it is, the forms it could take, the technologies that enable it, its potential applications, and the likelihood and timeline of its emergence. It also explores the possible societal impacts of the metaverse, including its benefits and risks, and conditions under which the metaverse could be developed and governed responsibly. Such an assessment is vitally important, even in the absence of certainty about the metaverse's emergence, as it allows stakeholders to anticipate developments, make informed decisions and, should the metaverse materialize, help shape its evolution in ways that serve the public good and minimize harm.

This book is intended for anyone seeking a deeper understanding of the metaverse—its possible emergence, potential benefits, risks, and ethical and societal implications. It is especially relevant for those involved in its research, development, or governance, including technologists, designers, policymakers, regulators, and civil society actors. Beyond offering insights, the book introduces practical tools and frameworks to support the responsible development, deployment, and oversight of the metaverse.

If the metaverse were to emerge as a mass medium, its impact could be as great as that of the internet, if not greater. The internet has fundamentally transformed how people communicate, access information, conduct business, and engage in social, political, and cultural life. The same could be true for the metaverse. The adoption of the internet has, however, come with great cost as well. The internet has enabled the spread of misinformation, large-scale surveillance, loss of privacy, social polarization, and new forms of exploitation like cyberbullying and digital addiction. Its control by a few dominant tech companies also raises concerns about accountability, censorship, and inequality—making it one of the most impactful and contested technologies of our time. Some of these harms might have been reduced through earlier intervention, better oversight, and more thoughtful planning. The metaverse offers a second chance to get it right—we should not waste it.

The structure of the book is as follows. Chapter 2 addresses the question, "What is the metaverse?" While public understanding of the concept is often vague or confused, there is considerable agreement among experts about the core features of the metaverse, its enabling technologies, and its potential applications. At the same time, important disagreements remain—these will be explored in detail. The chapter also introduces a typology of metaverse systems, clarifying different subcategories and forms the metaverse may take.

Chapter 3 examines the question, "Will there be a metaverse?" It analyzes both the driving forces behind the metaverse's development and the barriers that could hinder its emergence as a mass medium. The chapter argues that the metaverse is supported by strong drivers, including its potential to offer benefits that exceed those of the current internet and other digital technologies. There is also a compelling business case—assuming key obstacles can be addressed. Moreover, the metaverse aligns with broader technological, social, and economic trends, suggesting its development would not be an isolated phenomenon. At the same time, the chapter considers significant barriers that could limit the metaverse to niche applications or prevent its realization altogether. The discussion concludes with a balanced evaluation of its prospects for becoming a widely adopted platform.

Chapters 4 and 5 examine the potential impacts of the metaverse on individuals and society, assuming it evolves into a widely adopted medium. Building on Chapter 3's discussion of potential benefits, these chapters focus on key risks—highlighting potential threats to user well-being, individual rights, the public good, and environmental sustainability. Alongside this critical evaluation of how the metaverse could reshape core aspects of human and societal life, the chapters also offer proposals for how key stakeholders can anticipate and mitigate these risks.

Chapter 4 examines the potential implications of the metaverse for individual rights, with a focus on security, freedom, privacy, equality, and property rights. Special attention is given to challenges that are distinctive to immersive virtual environments. The chapter explores issues such as surveillance of virtual spaces and user bodies, virtual harassment and sexual misconduct, manipulation through immersive advertising and AI bots, censorship of oral speech and gestures, and theft, fraud, and exploitation involving digital and virtual property. The chapter also explores strategies for mitigating these risks and reinforcing the protection of individual rights within the metaverse.

Chapter 5 considers potential benefits and risks related to individual well-being, social practices and institutions, and the environment. It begins by examining how the metaverse can both enhance and threaten core aspects of well-being. While it may support meaningful experiences and relationships, it also poses risks to mental health, such as addiction, social isolation, and harmful self-comparison. Social well-being may be enriched but is also vulnerable to challenges like virtual infidelity and emotionally manipulative bots. Developers and operators have a responsibility to assess these effects and prioritize user well-being in platform design. This is followed by a broader analysis of risks and benefits for society, including implications for social institutions and social goods and practices. The chapter concludes with a brief assessment of the metaverse's environmental risks and sustainability potential.

For the analysis in Chaps. 3, 4, and 5, this book adopts the Anticipatory Technology Ethics (ATE) approach (Brey 2012; Brey et al. 2022), a well-established method for analyzing the potential benefits, risks, and social and ethical consequences of emerging technologies through foresight. ATE employs tools like trend analysis, scenario building, and expert consultation to explore plausible futures—encompassing potential products, uses, governance models, and societal impacts—which are then evaluated in terms of what is considered right, just, good, or responsible within those futures. Foresight in ATE is not predictive but explores multiple plausible outcomes. Assessments are conditional: if a certain technology with specific features emerges, then social consequences and ethical issues may arise. This helps developers and stakeholders anticipate and address consequences and risks, guiding decisions on which futures to pursue or avoid.

Chapter 6, finally, addresses the challenge of ensuring the metaverse is developed and governed responsibly. Given the metaverse's unique nature and potential for both benefit and harm, the chapter asks: What frameworks, practices, and responsibilities are needed to guide its development and operation in ways that respect rights, promote the public good, and reduce risks? We begin by outlining a general multistakeholder framework for the responsible governance of the metaverse. This is followed by a more focused examination of two key areas: the responsible governance of metaverse development and of metaverse operations. For each, we identify relevant stakeholders and explore practical instruments and strategies to guide ethical and responsible conduct. An annex to the chapter provides a proposed set of ethics guidelines for the metaverse's development, deployment, and use.

References

Brey, Philip. 2012. Anticipatory Ethics for Emerging Technologies. *Nanoethics* 6 (1): 1–13.
Brey, P., King, O., Jansen, P., Dainow, B., Erden, Y. J., Rodrigues, R., Resseguier, A., Diez Rituerto, M., Hatzakis, T., & Matar, A. (2022). Generalised methodology for ethical assessment of emerging technologies (Deliverable D6.1 of the SIENNA project, Version 2.1). Zenodo. https://doi.org/10.5281/zenodo.7266895

Chapter 2
What Is the Metaverse?

> "A speech with magical force. Nowadays, people don't believe in these kinds of things. Except in the Metaverse, that is, where magic is possible. The Metaverse is a fictional structure made out of code. And code is just a form of speech—the form that computers understand."—Neal Stephenson, Snow Crash

Abstract This chapter of the book *The Metaverse: A Critical Assessment* aims to answer the question what the metaverse is or could be. It first investigates this question through a history of the metaverse as an evolving concept and ongoing technological pursuit. Next, it explores the concept of the metaverse through key characteristics, such as immersiveness, persistence, interoperability, and multifunctionality, examining what they entail and how essential they are to defining the metaverse. Third, it examines the key enabling technologies driving its development and use, including virtual and augmented reality, 3D modeling and animation, artificial intelligence, blockchain, Internet of Things and foundational infrastructure systems. The chapter concludes by proposing a definition of the metaverse, informed by the preceding discussion, while recognizing ongoing ambiguities in the terminology.

Keywords Metaverse · Neal Stephenson · Meta · Virtual reality · Augmented reality · Artificial intelligence · Internet of things · Immersiveness · Persistence · Interoperability

---

© The Author(s), under exclusive license to Springer Nature Switzerland AG 2025
P. Brey, *The Metaverse: A Critical Assessment*, SpringerBriefs in Ethics, https://doi.org/10.1007/978-3-031-93471-1_2

2.1 A Short History of the Metaverse, from *Snow Crash* to the Present Day

The breakthrough year for the metaverse was 2021. It all started with an announcement of Facebook Inc. and its CEO, Mark Zuckerberg. On October 28, 2021, Zuckerberg made a public announcement at a high-profile company event, in which he laid out a new central mission for his company, which was the development of the metaverse. He portrayed it as the "next chapter for the internet" (Zuckerberg 2021a), as a new, embodied type of internet in which people no longer experience and interact through screens but have an embodied presence in an interactive environment. As Zuckerberg stated: "Screens just can't convey the full range of human expression and connection. They can't deliver that deep feeling of presence, but the next version of the internet can. That's what we should be working towards." (Zuckerberg 2021b).

It was also announced that the company's name would be rebranded to "Meta", to symbolize the major reorientation of the company. The company event also contained demonstrations of metaverse technologies that the company had produced so far. In an October 17, 2021, press release, Facebook had already announced its plans to help build the metaverse and announced plans to create 10,000 new high-skilled jobs within the European Union over the next 5 years (Clegg and Olivan 2021). In this press release, the metaverse was described as "a new phase of interconnected virtual experiences using technologies like virtual and augmented reality. At its heart is the idea that by creating a greater sense of "virtual presence," interacting online can become much closer to the experience of interacting in person."

Meta's announcement resulted in very extensive media coverage. This should not be surprising. One of the largest tech companies in the world, with a market capitalization of US$922 billion, declared that the future of the Internet was called the metaverse, and that it was rebranding itself and directing all its efforts to making this new incarnation of the Internet a reality. Suddenly, everyone was talking about the metaverse, a concept that few had even heard of before Meta's announcement. Everybody was asking: What is the metaverse? Is it the future of the Internet? Which companies are best positioned to make it happen? What will happen next? The day after Meta's October 17 press release, Google trends show a huge spike in interest in the term "metaverse". Searches for the term "metaverse", saw a 50-fold increase compared to average of the year before. In the 12 months after Meta's announcement, the word "metaverse" is mentioned on more than 56 million new web pages (Google Trends 2022).[1]

---

[1] Six months later, in April 2022, the interest was still more than ten times what it was before Meta's announcements. Google search lists 81.700.000 results for the search term "metaverse" in the period 10/17/2011 to 10/16/2021, a ten-year period, and 22.700.000 results for the period 10/17/2021 to 4/16/2022, a six month period, indicating an increase in the growth rate of pages mentioning the metaverse of 455% during that six-month period (Google Trends 2022).

By late 2021, the concept of the metaverse had entered the public consciousness at a large scale, and it had become strongly associated with Meta. However, following Meta's announcement, many other large tech companies also went on record about their plans for the future. As it turned out, many other tech companies were also developing technologies that are key to the metaverse, and many of them were also developing metaverse strategies. The media saw a race for the development of the metaverse, and debated which companies were best positioned to win it. It was reported that next to Meta, major companies like Google, Microsoft, NVIDIA, Unity Software, Roblox, and Epic Games had invested heavily in hardware and software required for the metaverse and were all well-positioned for this supposed race. While this was already happening before Meta's announcement, the explosion of interest in the metaverse certainly inspired organizations to allocate additional resources to it. In 2021 and 2022, hundreds of new metaverse projects and initiatives were started by companies and research institutions. Consultancy firms issued business reports claiming a trillion-dollar market for the metaverse within a decade. Metaverse exchange-traded funds (ETFs) became a popular investment, as did metaverse cryptocurrencies.

For many, the metaverse was a novel idea, an invention of Zuckerberg and his social media company. But the concept of the metaverse already had a decades long history before Zuckerberg's announcement. Let us trace this history and also consider what happened after Zuckerberg's announcement.

2.1.1 Neal Stephenson's Metaverse

The original concept of the metaverse was introduced by Neal Stephenson in his 1992 novel *Snow Crash*, in which the Metaverse (with capital "M") is depicted as a hyper-realistic future virtual reality environment that users can access at any time and in which many spend a large portion of their lives. The Metaverse in *Snow Crash* is a fully immersive, unified, and persistent virtual reality environment that can be accessed through virtual reality goggles and headsets. Users appear as avatars–digital representations of the user or of a character played by the user–and interact with each other just as they would in the real world. Stephenson's Metaverse consists of a simulated sphere that is larger than the planet Earth, though only a fraction of it is used, in the form of a sprawling city organized around a roadway called the Street. Corporations and individuals develop and purchase real estate in the Metaverse. In the city, one can find shops, offices, entertainment complexes and other real estate. There is no teletransportation in the Metaverse; avatars move around on foot, by motorcycle or car, or through public transit. The Metaverse is owned and operated by a single organization, called the Global Multimedia Protocol Group.

When Neal Stephenson introduced this concept, both the Internet and virtual reality (VR) had just become familiar to the general public. The Internet was in the early stages of adoption, and though the general public was not very familiar with it, tech enthusiasts had knowledge of it. VR technology emerged in the 1980s and by 1992, it started to capture the popular imagination. As Stephenson states in an acknowledgments section, his Metaverse was directly inspired by the emergence of VR, but he did not want to use the term "virtual reality", or similar ones that were used in the computer graphics community, because he considered them to be too awkward. Hence, he invented a new term in his novel, that of the Metaverse.

2.1.2 The Metaverse in Fiction

Stephenson coined the term "metaverse", but he is one of many science fiction authors who explore the idea of VR. Even before it existed as a technology, writers were already exploring the idea of a simulated reality as early as the 1930s. Laurence Manning's 1933 series of short stories titled "The Man Who Awoke"—later a novel—features a machine to which people connect, which replaces all their senses with electrical impulses and allows them to live a virtual life chosen by them. Stanley G. Weinbaum published a short story in 1935, Pygmalion's Spectacles, that featured goggles that let the wearer experience a fictional world through holographics, smell, taste, and touch.

Since the breakthrough of VR in the 1980s, it is increasingly featured in fiction. With the advent of the Internet in the 1990s, the topic of shared VR, in particular, became popular in fiction. Since the 1980s, many movies have come out that feature shared VR, including *Tron* (1982), *The Lawnmower Man* (1992), *The Matrix* franchise (1999–2021), *Total Recall* (1990), *Johnny Mnemonic* (1995), *EXistenZ* (1999), *Avatar* (2008), *Inception* (2010) and *Ready Player One* (2018). A number of television series also prominently feature VR, including *Star Trek: The Next Generation* (1987–1994), which featured a VR platform called the holodeck, *Black Mirror* (2011–2022), *Upload* (2020–2022), and *The Peripheral* (2022).

2.1.3 Gaming, the Internet and Shared Virtual Spaces

Through these works of fictions, the idea of a metaverse, understood as a shared, multi-purpose VR environment, had become implanted in the public consciousness, even though most novels and movies do not refer to such environments as metaverses. But it is not only works of fiction that stirred the public imagination. In the 1980s and 1990s, people became users of personal computers (PCs) and the Internet. PCs do not present users with immersive VR environments, but they present them with computer-generated environments: graphical user interfaces, gaming worlds, and simulations of various kinds. It is only a small step, conceptually, from such computer-generated environments to an immersive virtual environment.

The Internet moreover introduced the idea of shared computer-generated environments: shared web pages, chatrooms, bulletin boards, game environments, and other structures sustained by networked computers. Such online environments were also called "cyberspace": the shared, digital virtual space that constitutes the Internet. The metaverse is basically a combination of computer simulation, Internet connectivity, and immersive VR. People had already become familiar with the first two through their experience with PCs and the Internet; they knew of the existence of VR through news reports, and they knew of their combination into shared VR environments from novels and movies. So, by the 1990s, the idea of a metaverse as something that might someday come into existence was already embedded in the collective mindset.

2.1.4 The Metaverse as a Developer Concept

Technology developers started referring to the metaverse as a developer concept in the early 2000s. The idea of a metaverse was included in visions of a future Internet which included a 3D Web (World Wide Web): web environments that were three- rather than two-dimensional. In 2006, the Metaverse Roadmap Project was initiated by the Acceleration Studies Foundation, a US-based educational nonprofit organization. This project had the purpose of supporting the exploration of the virtual and 3D future of the Web. A Metaverse Roadmap Summit with futurists and developers was held in 2006, and a Metaverse Roadmap report was published in 2007 (Smart et al. 2007). The report explicitly referenced Neal Stephenson's Metaverse as an inspiration. It presented a variety of definitions of the metaverse, including definitions such as "The convergence of 1) virtually enhanced physical reality and 2) physically persistent virtual space", "A shared virtual social space with 3D capacity", "Our collective online shared space", and "A virtually-enabled physical world".

The report claimed that many of the Internet activities associated with the 2D Web would migrate to the 3D spaces of the metaverse. This does not mean that the entire Web would become 3D, but that it would become 3D-enabled. Moreover, it also does not mean that there would be one single metaverse, but rather that there would be "multiple mutually-reinforcing ways in which virtualization and 3D web tools and objects are being embedded everywhere in our environment and becoming persistent features of our lives". Importantly, this first conception of the metaverse as a developer concept evolves beyond Stephenson's concept in a number of ways. First of all, it breaks away from Stephenson's single, unified Metaverse. It also emphasizes a role for augmented reality (AR) as a central metaverse technology, which was not the case in Stephenson's original vision. In addition, it gives a large role to 3D visualization of Web content, which may or may not be accessed through VR displays. It moreover emphasizes the continuity between virtual worlds and online shared spaces, and it positions the metaverse as the future of the Internet for all its major uses.

2.1.5 The Rise of Metaverse Technologies

The 2000s and 2010s saw few efforts to develop an actual metaverse or 3D Web, but there were important developments in technologies and applications that are crucial to the development of the metaverse, including 3D modeling and animation, VR and AR, online virtual worlds, blockchain, cloud computing, and 5G networks.

In the 2000s, massively multiplayer online games (MMO), such as World of Warcraft and RuneScape, had a breakthrough. This marked the first time that users, represented by avatars, could interact with other users in large-scale, persistent 3D virtual environments. One virtual world, Second Life (2003), stood out because it was not really a game but a rather simulation of real life. It could be used for entertainment but also for more serious purposes like commerce, education, training, social networking, and community building. Second Life also introduced the concept of user-generated content and digital economies, enabling players to create, buy, and sell virtual goods. Second life is now known as a proto-metaverse or early metaverse. It did not offer integration with VR technology, but it was a persistent, massively multi-user virtual environment with a wide range of applications, just like Stephenson's Metaverse.

Another key development was the maturation of VR, which achieved a significant breakthrough in the consumer market during the late 2010s. For the first time, consumers gained access to affordable, high-resolution VR systems for gaming, social VR, and, to some extent, professional applications. Millions of units were sold, with the market being dominated by the Oculus Rift (developed by Meta, then Facebook), the HTC Vive, and the PlayStation VR.

The late 2010s also saw AR gaining mainstream traction, mainly through smartphones and tablets. AR games like *Pokémon GO* became popular, as did social media filters, virtual try-ons for shopping, AR educational tools, and AR navigation in Google Maps. 3D modeling and animation reached the point of photorealism by the late 2010s, and advanced graphics cards allowed for real-time rendering of highly realistic 3D environments. Another relevant development was the introduction in 2008 and 2009 of blockchain and cryptocurrency, and its integration into virtual worlds such as Decentraland (2020) and The Sandbox (2021). In addition, the emergence of 5G and the entrenchment of cloud computing in the late 2010s boosted VR with high-speed streaming and cloud-based rendering, reducing user hardware demands.

2.1.6 Mark Zuckerberg's Announcement and the Metaverse Initiative Boom

With the technological developments of the 2000s and 2010s in mind, the concept of building a metaverse was not as far-fetched as it initially sounded to many. Tech companies were already speculating on the future of human-computer interfaces as

2.1 A Short History of the Metaverse, from *Snow Crash* to the Present Day

well as on the future of the Internet. Would we always interact with computers through a screen and a keyboard, or might we do away with them and interact directly with digital objects through our gestures, body positioning, and speech, as virtual and augmented reality promised? If so, could we envision a future Internet which was presented to us in 3D rather than in 2D, featuring interactive environments and objects that mirror the physical spaces and items we naturally interact with? The technologies that could make this possible already seemed to be there, even though some were awaiting further development and maturation.

Let us consider these developments in more detail, beginning with the actions and strategies of tech companies. There is a fair amount of consensus on the technologies needed to power the metaverse. They are 3D modeling and animation, VR, augmented and mixed reality, blockchain, artificial intelligence (AI), Internet of Things (IoT), and various infrastructure technologies that include 5G, 6G, next-generation CPUs and GPUs, sensors and actuators, and cloud and edge computing. These enabling technologies are discussed in detail later in this chapter. As it turns out, in all these technology domains, companies had been making serious investments for years before Meta's announcement.

None of these technologies, though, are exclusively metaverse technologies. They all have a market status of their own. However, the metaverse buzz that ensued in 2021 has started affecting their development in several ways. First, tech companies that have started to include the metaverse in their business strategy were incentivized to make additional investments in these technologies. Second, tech companies that develop these technologies wanted to ensure that they will be developed in a way that supports integration with other metaverse technologies. These developments, in turn, could bring additional investments to these firms from investors who believe in the metaverse as a business opportunity.

It is noteworthy, in this context, that most large tech companies have been investing in what are perhaps the two signature technologies for the metaverse, VR and augmented and mixed reality (MR). Meta was the undisputed market leader in 2022 in VR hardware with its Quest headsets. Apple has been developing its own MR headset, the Apple Vision Pro, which was released 2024, and had also invested in AR applications. NVIDIA was heavily invested in VR and AR through high-performance GPUs, cloud-based streaming with CloudXR, and the Omniverse, a platform for real-time 3D collaboration, simulation, and digital twin creation across industries. Google had released low-cost VR headset and smart glasses and had made serious investments in AR. Microsoft invested in an advanced MR headset, HoloLens 2, that was released in 2019 but decided shortly thereafter to concentrate solely on VR/AR software and partnerships.

Many companies that were involved in the development of the aforementioned key technologies for the metaverse established metaverse strategies and initiated metaverse projects in the early 2020s. For example, NVIDIA, the leading manufacturer of graphics cards, stated in 2022 that the metaverse was central to its business strategy and boasted its Omniverse as a real-time graphics collaboration platform for the visual effects and digital twin industrial simulation industries (Takahashi, 2022). Microsoft affirmed in 2021 and 2022 that it was strongly committed to the development of the metaverse, especially through VR and AR software, and it has a particular focus what it calls the industrial metaverse (Grant, 2022). Also, in 2021, Disney announced plans to develop a metaverse theme park (Milmo, 2021). Some non-tech companies that are potential users of the metaverse also announced metaverse strategies and partnerships, including IKEA, Gucci, JP Morgan, and Lego. In China, some of the biggest tech companies, including Tencent, Huawei, Alibaba, and Baidu, also initiated metaverse strategies in the early 2020s. Many companies also featured their own proto-metaverses in the early 2020s, virtual worlds like Meta's Horizon Worlds, Baidu's XiRang, Decentraland Foundation's Decentraland, Pixowl's The Sandbox, and Roblox Corporation's Roblox.

In the early 2020s, several countries also announced national strategies for the metaverse. Among these, the most important is China. In its 2021 five-year plan, China identified key metaverse technologies like VR, AR, and blockchain as central to the development of its digital economy. In 2022, the Metaverse Industry Special Committee was established as a committee of the China Computer Industry Association, under the supervision of the Ministry of Industry and Information Technology (MIIT). It counted more than 150 companies as founding members. Its aim is to lead and drive the development of the industry, which includes taking the lead in drafting industry standards and formulating industry plans. It also includes driving the development of various specialized metaverses, among which are an industrial metaverse, health metaverse, life metaverse, medical metaverse, architecture metaverse, and business metaverse. In 2022, the MIIT also announced plans to cultivate 3000 startups in the fields of metaverse, blockchain, and artificial intelligence (Fabernovel 2022). Within the same year, roughly 1500 Chinese companies had already applied for trademarks related to the concept of the metaverse (Che 2022). Other countries that established national metaverse strategies in the early 2020s include South Korea and the United Arab Emirates. The European Union also announced plans for a metaverse strategy in 2022 (European Commission 2022; Bezmalinovic 2022). At the time, it already had developed economic strategies for many of its key technologies, including VR/AR, blockchain, and AI.

The year 2022 also saw the establishment of the Metaverse Standards Forum, an international organization that aims "to encourage and enable the timely development of open interoperability standards essential to an open and inclusive metaverse" (Metaverse Standards Forum n.d.). Within one month of its establishment, on June 21, 2022, it already had over a thousand organizational members, which had grown to almost 2400 a year later. Members include tech giants Microsoft, Google, Samsung, Meta, Huawei, Baidu, NVIDIA, and Intel, as well as the World Wide Web Consortium, the XR Association, and the Industry IoT Consortium, among many others.

On May 25, 2022, the World Economic Forum announced the formation of Defining and Building the Metaverse, an international initiative focused on the safe, secure, interoperable, and inclusive development and governance of the metaverse as well as on the strengthening of economic and social value creation opportunities for the metaverse (World Economic Forum 2022). The initiative was supported by over a hundred organizations, including Meta, Microsoft, the Internet Watch Foundation, the United Nations Office of Counterterrorism, the International Telecommunication Union, and several national governments.

While not all major tech companies had joined the metaverse bandwagon to the same degree, and not all liked and used the label "metaverse", by late 2022, development of the metaverse had become a serious undertaking across the world, supported by many major tech companies, as well as by many universities and research institutes, governments and intergovernmental organizations.

2.1.7 *Cooling-off and Realignment*

By 2023, the initial hype around the metaverse had resided, and a cooling-off period followed. Zuckerberg's announcement and overhyped media coverage made it seem as if an advanced, all-encompassing virtual world was just around the corner, but practical applications fell short, leaving users and investors disappointed. It became clear that it would still take many years before a metaverse could be built that would live up to the hype, and some were starting to believe that the whole idea was misguided. Apple's introduction of the Apple Vision Pro in 2024, the most advanced MR headset yet, did not help: Apply spuriously avoided any reference to the metaverse, or even to VR and AR, and instead introduced the concept of spatial computing.

The interest in the metaverse was also seriously affected by the increasing interest in another technology that started being hyped in the early 2020s: artificial intelligence (AI). AI was seen as a more mature technology, with immediate benefits for businesses and consumers. The breakthrough of generative AI in 2023 sealed the deal: tech companies, investors, and the general public jumped massively on the AI bandwagon, and interest in and investments for the metaverse was reduced as a result. By 2024, news reports started to appear with headings like "What happened to the metaverse?" and "Is the metaverse dead?" (Tess 2024; Sharma 2024). The story in the news was that Zuckerberg's metaverse had failed, and that either it was a bad idea from the start or was still many years away from realization.

Although the hype surrounding the metaverse has resided, many industry insiders believe that the metaverse is far from over. They still believe that VR and AR will play a pivotal role in the future of human-computer interaction, with much of computing occurring in immersive virtual and augmented environments, often online and within social contexts. Whether one believes in this or not, it is a fact that there are still serious efforts underway to develop key technologies that are needed for the metaverse, such as VR and AR technology, 3D modeling and animation, blockchain, and AI.

Even more so, there are still many collaborative efforts in the tech industry that take place under the metaverse banner. Internationally, many initiatives related to the metaverse are coordinated under the umbrella of the Metaverse Standards Forum, which brings together leading technology companies, developers, and organizations to establish interoperability standards and foster a cohesive, accessible metaverse ecosystem that operates seamlessly on a global scale. China, in addition, remains firmly committed to the metaverse. In September 2023, the Chinese government announced its Three-Year Action Plan for the Industrial Innovation and

Development of the Metaverse (2023–2025), with the aim of establishing industrial clusters to develop the next-generation, 3D immersive internet, with a special focus on industrial applications (the industrial metaverse) (Interesse 2023) It subsequently also created a large working group of tech leaders and university scholars to develop industry standards for the metaverse (Global Times 2024). Thus, even though the metaverse is on life support in the public eye, the reality in the tech world is different, with many companies still committed to both the technologies and the ideas behind the metaverse.

2.2 A Framework for Understanding the Metaverse

For members of the general public, the metaverse has remained a vague concept. What, exactly, is a metaverse, and what is it for? For many, this is still unclear. Within the tech community, there is greater clarity and consensus, though some points remain contested. The metaverse, it is usually agreed, consists of one or more three-dimensional, computer-generated, persistent virtual worlds. These virtual worlds are used by a large community of users, who are represented by avatars, and who interact with the world, the shared virtual objects in it, and other users. The virtual worlds are moreover accessed through VR and AR technology, allowing for immersive interaction with these worlds. It is moreover agreed that these worlds are multifunctional: they may be used for gaming and entertainment, but also for socializing, commerce, education, virtual work, and healthcare. To the extent that the metaverse consists of multiple virtual worlds, these worlds are interconnected and interoperable, making for a seamless connection between them.

We will now consider in more detail, key dimensions of the metaverse on which there is broad agreement, and second, aspects on which there is disagreement. We will start with the notions of a virtual world and with a key disagreement: the role of AR in the metaverse. We will then explore the dimensions of.

2.2.1 *Virtual Worlds: Unity, Interoperability, and Persistence*

As a core element, it is often agreed, the metaverse includes one or more virtual worlds. A virtual world is a computer-generated environment where users can interact with digital objects, settings, and often other participants in real-time. While virtual worlds can in principle be text-based or 2D simulations, the virtual worlds of the metaverse are projected to be 3D graphically rendered environments. 3D virtual worlds emerged in the 1990s, in gaming and entertainment. The current virtual worlds of Minecraft, Fortnite, and Roblox have tens of millions of users.

A point of contention has been whether the metaverse should consist of one single, unified world or multiple interconnected worlds. A unified metaverse is a virtual world that is defined within a single three-dimensional Euclidean space within which all objects and structures of the metaverse are defined. Any object or structure of the metaverse has its own coordinates in this space, and users can travel, in principle, to any other place in the metaverse by crossing the relevant distance. This is the classical metaverse as found in Stephenson's *Snow Crash*. Most experts agree, however, that such a unified metaverse is not likely to emerge. Developers are now all building separate virtual worlds, which gives them the advantage of being able to build their own world according to the specifications that they desire. Working within a single spatial grid at essentially the same virtual world would require that developers adapt to a vast number of standards, business models, and creative ideas. A unified virtual world also means that there is only one VR, which negates the idea of virtual worlds as alternate and potentially incompatible realities, as works of fiction are.

An alternative to a unified metaverse is a plurality of disjointed metaverses. In this scenario, different companies and creator communities build different metaverses, resulting in a multiplicity of metaverses with their own user base and their own unique features. Avatars and digital assets cannot be transported between these metaverses. This scenario would mimic that of the current Internet, in which there are multiple social media sites, online game worlds, and non-gaming virtual worlds, and people cannot easily transport their identities, data, and assets between them.

Metaverse advocates are converging on a third option: the metaverse as a network of interconnected virtual worlds. What does it mean for virtual worlds to be interconnected? The key notion that has been proposed for this is that of *interoperability*. In the context of the metaverse, interoperability is commonly understood as virtual worlds adhering to similar standards so as to make information exchange between them possible, especially regarding the ability of users to transport their identity, data, and digital assets between virtual worlds.[2] Interoperability is seen as a highly desirable feature for the metaverse, as it allows users to seamlessly travel between virtual worlds and to preserve and use their digital assets across them. It would also allow users of different virtual worlds to interact and collaborate with each other across worlds. In this way, disjointed virtual worlds can still be experienced as a part of a larger universe that is navigated and interacted with by the users.

Persistence is another quality often attributed to virtual worlds of the metaverse. A persistent virtual world is a world that, as defined by virtual world expert Bartle, "continues to exist and develop internally (…) even when there are no people interacting with it" (Bartle 2003, 2). Persistence may also be a quality that is attributed to avatars and virtual objects. Persistence makes virtual worlds more believable and

---

[2] This interpretation builds on meanings of the term in engineering and computer science. In engineering, interoperability is understood as a characteristic of a product or system to work with other products or systems, and in computer science, it is understood more specifically as the ability of computer systems and software to exchange information and use the information that is exchanged.

realistic, it guarantees their continued availability to users, helps build community, and invites long-term investments in virtual worlds. For certain applications, however, world persistence simply does not seem to add much value. Virtual business meetings, for example, only require generic rooms with generic furniture, and persistence would add little value. Persistence may become a feature of metaverse environments when it has added value, such as for shared public spaces and digital twins used for real-time simulations in science, engineering, and education, but likely not for virtual environments used for one-time interactions, such as business meetings, event spaces, and training simulations.

2.2.2 Extension to Augmented Reality

In earlier definitions of the metaverse, it was conceived of as an interconnected network of 3D immersive virtual worlds that are accessed through VR technology. In many later definitions, it is held that the metaverse consists of both fully virtual worlds and MR environments, which are environments consisting of both virtual and physical elements, such as a construction site in which a 3D simulation of a building is projected to help engineers align work with the design. The metaverse is moreover accessed through both VR and MR/AR technology. Specifically, virtual environments are interacted with through VR interfaces, and MR environments are engaged with through mixed and augmented reality interfaces. This sometimes results in a distinction between two versions of the metaverse: the VR metaverse, which is based on virtual worlds and VR technology, and the AR metaverse, which is defined over MR environments and augmented and mixed reality technology.

A problem for this extension of the classical conception of the metaverse is that MR environments are often not social, not persistent, and not interconnected, all three qualities that have long been strongly associated with the concept of the metaverse. They are often not social because they are often designed to provide personalized information to individuals to inform and guide them in specific tasks, presenting virtual objects and information relevant for the task. They are often not persistent because the virtual elements are often added for temporary use. And they are often not interconnected with other mixed or virtual environments because of their individual, temporary use.

So how to make sense of the role of mixed and augmented reality in the metaverse? The notion of an AR metaverse on its own does not seem to have much content, for the reasons given. But one can certainly extend the notion of a VR metaverse to also include those mixed realities that are social, interconnected, and to some degree persistent. Specifically, the metaverse could be defined to include shared, interoperable virtual objects and MR spaces, such as the virtual building on the construction site referred to earlier. It would not include personalized, temporary MR environments. This implies that the role of AR in the metaverse is somewhat limited, at least currently, as many current applications of AR are personalized and temporary.

2.2.3 Immersiveness

Immersion is the feeling of being present in a virtual or media-generated world. Immersiveness is the quality or degree to which a VR system or media platform is immersive. Immersiveness is generally considered to be a key quality of the metaverse, since it enhances qualities of interaction with that world and with others. Although immersiveness is to some extent in the eye of the beholder, 2D media are generally considered to have low immersiveness and 3D interactive media high immersiveness. Immersiveness is moreover influenced by features such as sensory fidelity, interactivity, and realism. VR and AR are considered to be immersive media, since they immerse the user in a 3D environment that is partially or wholly computer generated, with realism being added by the environment's seamless response to the user's body position and movement.

It is helpful to distinguish different levels of immersiveness for VR, AR, and the metaverse. There is no accepted standard for distinguishing different levels, so this is one of the first attempt to define them:

- Level 0. *Non-immersive virtual worlds and environments.* These are 3D computer-generated virtual worlds that are experienced through the screen of PCs and mobile devices, and that either use a regular interface (level 0a) or make use of some positioning and motion tracking or stereoscopy (level 0b).
- Level 1. *The audiovisual metaverse.* These are virtual worlds that are provided through visual and aural stimuli, and that involve either stereoscopic vision on desktop or mobile screen or 360-degree monoscopic vision through a headset or screen projections (level 1a); 360-degree stereoscopic vision and head, controller and hand tracking (level 1b); or 360-degree stereoscopic vision and full-body tracking (level 1c).
- Level 2. *The haptic metaverse.* These are virtual worlds rendered by means of visual, aural, tactile, and proprioceptive stimuli. They are assumed to have the qualities of the level 1b or 1c audiovisual metaverse, plus tactile and motion feedback. This can come in the form of haptic gloves (level 3a), haptic suits (level 3b), or haptics plus the additional use of treadmills and motion simulators (level 3c).
- Level 3. *The pan-sensory metaverse.* These are virtual worlds provided through all the senses: visual, aural, tactile, olfactory, and gustatory, plus our sensory systems for balance (vestibular system), movement (proprioception), and possibly even inner bodily states (interoception).
- Level 4. *The hyper-realistic metaverse.* These are virtual worlds based on advanced VR technology that has reached the point at which VR is indistinguishable from physical reality.

Immersiveness is a complex quality that is not easy to put on a linear scale. It is not only defined by the mentioned criteria, but also by the level of realism of the sensory stimuli and the seamless operation of the technology. Many developers of VR and AR technology have hyper-realism as a goal, including market leader Meta (Smith 2022). For auditory and visual stimuli, the technology is already nearing a state of hyper-realism.

2.2.4 Multifunctionality

Online platforms and software applications on the current Internet are usually specialized: they support one function, such as gaming, office work, or online shopping. The metaverse is often thought of as having many purposes and serving different needs. Second Life, the proto-metaverse discussed earlier in this chapter, is an open world that allows people to do many of the things that people can do in ordinary life (hence the name): socialize, go shopping, start a business, be a DJ in a virtual nightclub, explore the region, play a game, or attend a class. The metaverse is often envisioned to be similarly multifunctional, supporting many of the key practices and social and economic functions of the real world, such as socializing, entertainment, commerce, work, education, and healthcare.

An alternative, popular view advocates for building specialized metaverses (Lowry et al. 2024). The *industrial metaverse*, focused on manufacturing, has gained traction globally, with companies like Siemens partnering with NVIDIA and Microsoft with Kawasaki in 2022 (Cureton 2022; Kovach 2022). It centers around digital twins, virtual real-time replicas of production systems and products for planning, testing, monitoring, and maintenance. Similar initiatives also exist around the notion of an *enterprise metaverse*. Whereas the industrial metaverse simulates manufacturing processes, the enterprise metaverse, also called *business metaverse*, is focused on simulations that can help businesses grow, including remote collaboration, remote operations, meetings and events, training, digital marketing, retail, and the production of digital goods (Avanade 2022).

The industrial and enterprise metaverse are sometimes contrasted with the *consumer metaverse*, the metaverse for ordinary people, for socializing, gaming, entertainment, shopping, and similar pursuits. Game platforms like Roblox and Fortnite are sometimes called (proto) *gaming metaverses*, and Meta, a social media company, sometimes labels itself a "social metaverse company". Existing virtual worlds in the consumer space often have one or a couple of main functions, which can be gaming, socializing, content creation, virtual events, commerce, learning, or a combination of several of these. Another type of metaverse, the healthcare metaverse, is a virtual environment integrating advanced technologies to enhance medical services, education, and patient care through applications like telemedicine, immersive medical training, therapy, and digital twins for personalized treatment.

2.2.5 Openness and Decentralization

Developers and advocates across the spectrum support an open metaverse. However, they often mean different things with it. One point of agreement is that openness implies interoperability between different virtual worlds, which was already discussed. A second point of agreement is the use of open standards (Perri et al. 2023). This means that the technical specifications, protocols, and formats that facilitate communication, data processing, and content creation are nonproprietary and available for use by anyone. Open standards contribute to interoperability, but in addition they also ensure that not one company owns and controls the metaverse, and that a market is supported that is likely to be both collaborative and competitive and fair to different kinds of businesses. A third point of agreement is that the metaverse should be inclusive: open to all users and based on principles of fairness and equal opportunity.

Various additional characteristics have been proposed to be included in the definition of openness. These are characteristics that are intended to decentralize control, empower users, and further level the playing field for developers and creators beyond the big tech companies. In particular, openness has been associated with decentralization, community ownership, user governance, open source, self-sovereign identity (self-management of one's digital identity without dependency on third-party providers), strong privacy protections, and freedom of expression.

Decentralization is the key concept on this list. A centralized metaverse is one that is owned, operated, and governed by one or a handful of powerful companies. In such a metaverse, control is in the hands of one or a few players. A decentralized metaverse is usually defined to be included in a decentralized Internet (Web3). It is a metaverse that has collective ownership by its users, and decentralized operation and governance. Open standards, public blockchain, open source, and a creator economy would support such a decentralized metaverse.

The choice between a centralized and a decentralized metaverse has enormous implications: a centralized metaverse offers streamlined user experiences and efficient management, but this comes at the cost of reduced user control, monetization, and potential monopolization. A decentralized metaverse fosters greater user autonomy, data ownership, and interoperability but could face scalability and governance challenges.

2.2.6 Relation to the Internet and to the Physical World

There are three main positions on the metaverse's relationship to the Internet. The first sees the metaverse as the future Internet, fully replacing it. As VR pioneer Tony Parisi has stated in a well-known manifesto: "The Metaverse is the Internet, enhanced and upgraded to consistently deliver 3D content, spatially organized information and experiences, and real-time synchronous communication." (2021). The second views the metaverse as an expansion of the Internet, coexisting with the current 2D Web, as many users may still rely on simpler, screen-based interfaces for many everyday tasks, like looking up information or paying bills. The third position envisions the Internet evolving into Web3 or Web 3.0, integrating technologies like blockchain, the Semantic Web, and IoT, with the metaverse as just one part of this broader system. The outcome will likely depend on the evolution of human-computer interfaces—whether VR and AR fully replace screens, keyboards, and mice or continue to coexist alongside them. If AR and VR interfaces dominate, the metaverse could become synonymous with the Internet; otherwise, it will remain a part of it.

Another contested issue is the metaverse's relation to the physical world. The classical metaverse, as envisioned by Stephenson and early advocates, is a self-contained virtual world that has little interaction with the physical world. As argued, it is more likely that the metaverse will include both virtual and mixed physical-virtual environments, the latter being realized through augmented and mixed reality devices. As a consequence, the metaverse will be integrated in part with the physical environment. In addition, with the emergence of the IoT, a global network of Internet-connected devices, other mergers with the physical world arise. Most importantly, IoT is involved in the creation of digital twins, which are particularly important for metaverse applications in science, industry, healthcare, and education, and which have a physical component. The metaverse may also present virtual interfaces to operate IoT devices from inside the metaverse, which is another way in which the metaverse and the physical world can interact.

2.2.7 Degree of Social, Economic, and Political Organization

If the metaverse becomes a place where one can meet one's friends, go to a concert, go shopping, or participate in a lecture, and experience all these events in a way that is almost as real as in the physical world, and perhaps in some ways even better, then the metaverse may become a place where some people will spend a large part of their lives. This is even more the case for people who have employment in the metaverse, for example as store clerks, performers, educators, or tour guides. This raises the question whether a metaverse, particularly a multifunctional one in which many people "live" a large part of their lives, should not have a certain amount of social organization, including developed social institutions. Should it have its own economic system, currency, laws and regulations, and even a form of government? Should it have churches, schools, universities, banks, stores, community service organizations, a police force, and journalistic media?

Two perspectives address this question. The *metaverse subservience view* sees the metaverse as an extension of physical society, governed by existing institutions and supplemented by platform-specific rules. In contrast, the *metaverse self-governance view* argues that as metaverses grow and people spend more time in them, they will resemble cities or nations, requiring their own governments, policies, and institutions, especially given their likely transnational nature. For governance of the metaverse, it is important to consider stakeholder needs for institutionalization and social organization, taking into account their preferences and interests.

2.3 Key Metaverse Technologies

Even though there are different visions of what the metaverse is or should be, there turns out to be considerable agreement as to what technologies are needed for it. In this section, we will introduce each of these technologies and analyze their role in the metaverse.

2.3.1 3D Modeling and Animation

Metaverses are computer-generated worlds populated by avatars and objects that display animated behaviors and respond to user interaction. These worlds are created using 3D modeling and animation. *3D modeling* involves creating mathematical, three-dimensional representations of objects, which can be rendered into images or used in simulations, while *3D animation* defines motion patterns for lifelike movements, such as waves, flight, or human actions. Together with 3D sound, these techniques simulate highly realistic, interactive environments that respond to user input in real time using 3D engines. Together, 3D modeling and animation can be used to simulate highly realistic objects and environments in 3D, whether for use in movies, games, or professional simulations. For VR and AR applications, 3D modeling and animation needs to be specially tailored to ensure real-time rendering, immersive interaction, and precise spatial alignment, so as to render lifelike environments and objects that respond dynamically to users.

A practice in 3D animation and modeling that is of particular relevance to the metaverse is *3D reconstruction*. This is a type of 3D modeling which captures the shape and appearance of real objects. This requires detailed input from the source object, which is obtained through sensing techniques or 2D images. 3D reconstruction is used to create exact virtual replicas of unique objects, buildings, and scenes, and to add realism to virtual environments. Another relevant activity is the development of *digital twins*, virtual models of objects or systems that are modeled over time, with updates from real-time data from their physical equivalent. Digital twins are used for simulation, experimentation, planning, and decision-making in science, engineering, and medicine.

2.3.2 Virtual Reality

VR goes beyond 3D modeling and animation by providing an immersive, interactive experience that allows users to engage with computer-simulated environments in a way that feels real, potentially involving all five senses. VR intends to immerse the user into a virtual world of which they are a part, rather than to merely present them with a 3D rendered object or environment on a display. VR is a defining technology for the metaverse because it enables immersive, embodied experiences that allow users to interact with digital environments and each other as if physically present.

When it emerged as a viable technology in the 1980s, VR was marketed as consisting of input/output devices that covered three of the senses. Head-mounted displays (HMDs) offered surround stereoscopic vision and surround sound. Datagloves and datasuits tracked the positions and motions of body parts so as to instruct the computer to modify its output depending on the recorded positions. They could potentially also provide tactile feedback from sensed virtual objects. These original technologies have shaped what is still the most common understanding of "virtual reality": an immersive, interactive 3D computer-generated environment in which interaction takes place over multiple sensory channels. Central to the idea of VR is the idea of presence: being present in an environment rather than observing an environment from the outside.

Sherman and Craig (2003) have claimed that VR is defined by four key elements: a virtual world, immersion, sensory feedback, and interactivity. A *virtual world* is a basically a collection of 3D modelled objects in space, plus rules and relationships governing these objects. *Immersion* is the sensation of being present in an environment, rather than just observing it from the outside. *Sensory feedback* is the provision of sensory data about the environment, which is adapted in relation to the position and actions of the user. Finally, *interactivity* is the responsiveness of the virtual world to user actions; this includes the ability of the user to navigate the virtual worlds and to interact with objects, characters, and places. Immersion and interactivity are particularly central, creating a sense of presence by allowing users to feel fully surrounded by the virtual environment and engage with it using their entire body rather than just a keyboard and mouse.

Every VR system places somewhere on a scale from non-immersion to full immersion. Non-immersion is the condition in which one is interacting with a virtual environment indirectly, through devices like keyboards and mice, and perceiving it as an image encountered on a screen. Full immersion is a condition in which the virtual world is fully responsive to one's body, all five senses are stimulated by it from each direction in space, and one's perceptions of it are so realistic that one experiences a strong feeling of presence. Among the qualities that enhance the immersion of a VR system are realistic 3D modeling and animation, panoramic display covering the entire visual field, stereoscopic imaging, the use of 3D sound, interactivity of the virtual world in response to tracking of the position and motion of one's body, head, limbs, and eyes, the provision of tactile and kinesthetic feedback to one's body, and the use of digital scent technology and taste synthesizers to evoke smells and taste. Moreover, the maintenance of a high throughput rate for inputs and outputs and the minimization of lag, which is the delay between performance of an action and seeing the results of that action, are also important for realism.

Definitions of VR vary, from narrow interpretations requiring advanced setups like HMDs and motion tracking, to broader ones that include desktop or mobile systems where users interact with virtual worlds through avatars or devices like PCs, tablets, or smartphones, often referred to as non-immersive VR. For this book, VR is defined as requiring immersion and presence, characterized by sensory feedback relative to the user's body and the illusion of being within the environment, achieved through 360-degree vision, stereoscopy, or both. This minimal definition emphasizes interactive, 3D virtual environments that respond to user movements, typically experienced through VR headsets or stereoscopic setups.

VR has been steadily improving over the years, and there is now a multi-billion dollar commercial market for head-mounted displays (also called VR headsets) and VR accessories. Contemporary VR headsets have high framerates, limited lag, high resolutions, large fields of vision, and surround sound. They often have built-in abilities for tracking their own location and movement as well as those of any controllers that are used and, sometimes, also the hands of the users. They often come with VR controllers, most of them being hand controllers that provide button input and make use of motion tracking, gesture interfaces, and position tracking technology to track hands and fingers.

The VR accessories market is vast, with products enhancing immersion through tracking, feedback, and simulation. Key categories include VR trackers for full-body and object tracking, facial trackers for expressive avatars, and VR suits that offer full-body motion capture and haptic feedback like touch, pressure, and temperature. Some suits also track biometrics. Haptic gloves and shoes provide similar features, while teledildonics simulate remote sexual interaction. Additional accessories include VR treadmills for 360° movement, motion simulators for realistic vehicle sensations, and sensory devices like scent masks and taste stimulators. Physical tool replicas—like gun stocks, paddles, and simulators—further enrich the VR experience.

The future may bring a new way to interact with virtual environments: *brain-computer interfaces* (BCIs). These systems record and analyze brain signals, translating them into commands for output devices. BCIs rely on detecting intent from brain activity—such as the desire to move or control something. They could eventually replace traditional input devices in VR (Coogan and He 2018). Basic BCI-enabled VR headsets already exist. A more advanced concept involves not just reading brain signals but also stimulating neural tissue to create sensory experiences. This could enable highly realistic virtual environments via direct brain interfaces.

A key distinction in VR systems is between *single-user* and *multi-user VR*. Single-user VR involves just one person and is used for activities like training, education, or single-player games. Multi-user VR, or *shared* VR, connects several users over the Internet or a local network. A common form is *collaborative VR*, where remote users interact and work together in a virtual space. Multi-user VR is also used for socializing, gaming, shopping, and watching performances. In these environments—especially collaborative ones—communication is essential, so VR systems should support speech, gestures, movements, and facial expressions for avatars. The metaverse centrally involves multi-user VR.

In multi-user VR, as well as in some single-user VR systems, users appear as *avatars*: 3D characters whose movements are controlled by the user. Avatars can resemble the user but can also represent another person or a fantasy character. They can be viewed from a first- or third-person perspective and controlled via keyboard, gamepad, or, in advanced systems, through VR trackers that capture facial expressions and body movements. In shared virtual spaces like the metaverse, users interact with each other through these avatars.

2.3.3 Augmented and Mixed Reality

AR enhances our experience of the real world by overlaying computer-generated visuals, sounds, or other sensory input. Most AR systems focus on visual elements like text, textures, or 3D images. Unlike VR, which creates separate environments,

AR adds virtual elements to the real world, offering real-time information, enhancing object recognition, and supporting tasks like gaming, learning, and collaboration. AR can be experienced through a smartphone or tablet, through AR video projectors, a head-mounted display, dedicated glasses called smart glasses, or through contact lenses (in development). AR is currently a multibillion-dollar market.

Mobile AR is currently the most common form. It uses smartphones or tablets to overlay digital content onto the physical environment through the device's camera, sensors, and AR-enabled apps or websites. Examples include Snapchat filters and Pokémon GO. AR headsets and smart glasses are also rising in popularity. Since the commercial failure of Google Glass in 2015, smart glass technology has improved significantly, despite ongoing privacy concerns. Most models feature built-in cameras, microphones, speakers, and gesture controls, with processing handled by a connected smartphone or mobile device. Most smart glasses do not offer true AR. Instead, they display images, videos, or information in the user's visual field, typically triggered by voice commands or touch input through a connected smartphone, rather than responding to features of the surrounding environment. For smart glasses to provide genuine AR, they must be able to sense and understand the environment, placing virtual objects within it as if they were real. While such devices do exist, their capabilities are often limited, especially in terms of how complex or interactive the virtual elements can be.

More advanced are AR head-mounted displays, which, unlike most smart glasses, offer larger screens, better sensors, and built-in computing. Devices like Microsoft's HoloLens 2 can sense the environment and user in detail, projecting dynamic, interactive 3D objects. Some AR glasses now share similar features. Still in development, AR contact lenses—like Mojo Vision's prototype—aim to project visuals directly onto the retina using a micro-display, sensors, and micro-batteries.

AR has a wide range of applications. It serves as a powerful collaborative tool in fields like engineering, architecture, manufacturing, logistics, and urban planning by enabling interactive 3D models for analysis and design. In commerce, AR enhances online shopping by letting users view, test, and place virtual products in real-world settings. It also supports education and training through interactive visualizations and real-time information. In healthcare, AR aids in medical training, diagnosis, procedure planning, and therapy for cognitive impairments. It can enhance social interaction and is popular in gaming, with Pokémon Go as a major example. Other uses include navigation, broadcasting, sports, and sightseeing.

What does AR have to do with the metaverse? Using AR on a smartphone to project a couch into a room does not feel like entering the metaverse, often imagined as a shared, immersive virtual world. Still, many metaverse advocates now view AR as a key technology. VR and AR pioneer Louis Rosenberg (2022) proposes an *AR metaverse* instead of the classical *VR metaverse*, defining it as "the merger of real and virtual worlds into a single immersive and unified reality".

A problem with this proposal is that it deviates significantly from the popular idea of a metaverse as a shared, persistent, immersive virtual world. Most applications of AR involve only a temporary addition of virtual objects to the real world, they only have a single user or just a handful of users, and there is no

2.3 Key Metaverse Technologies

computer-generated world that can be navigated. Including these applications in a definition of the metaverse would stretch the notion too far. The metaverse is by definition (massively) multi-user and persistent. It therefore includes multi-user VR applications in which persistent virtual environments are generated. But it could also include multi-user VR, in which (unique) virtual objects and environments are generated and blended with physical space.

Alongside VR and AR, MR is often proposed as a third category of computer-supported reality and a key technology for the metaverse. However, the term is used inconsistently. Speicher et al. (2019) found at least six different meanings in expert interviews and literature. We will focus on the two most common.

First, Milgram and Kishino (1994) introduced MR as part of a reality-virtuality continuum, ranging from the physical to the virtual. By their definition, MR includes any blend of physical and virtual elements—such as AR and augmented virtuality (AV), where physical elements are integrated into a virtual environment in real time.

Second, MR is sometimes defined as an enhanced form of AR, situated between AR and AV. Sometimes, the term "AR+" or strong AR is used instead. Unlike "ordinary" AR, where virtual objects are overlaid and mostly passive, MR involves interactive, context-aware virtual elements that respond to both users and physical objects. This creates a seamless blend where the real and virtual can be hard to distinguish. However, this definition depends on a narrow view of AR—some definitions already include such interactivity, making MR redundant.

Given the ambiguity surrounding "mixed reality," we will avoid the term in this book. Instead, we will differentiate between weaker and stronger forms of AR, and use *"extended reality"* (XR) as an umbrella term covering VR, AR, AV, and MR.

2.3.4 Artificial Intelligence

AI is a field of computer science that began in the 1950s but saw major breakthroughs in the 2010s and 2020s. It focuses on creating systems that perform tasks requiring human intelligence, such as reasoning, problem-solving, planning, language use, image recognition, and learning. Research often targets specific abilities, with AI excelling in some areas and progressing in others. Today, AI is widely applied across society—from industry and healthcare to education and entertainment. Many AI applications rely on *big data* to identify patterns, make predictions, and suggest actions. A key driver of this is *machine learning*, a dominant subfield of AI that enables systems to learn from data and improve through feedback, making them increasingly adaptive and autonomous.

AI plays a key role in making the metaverse smarter, faster, and more realistic. It is essential at every layer—from infrastructure like 5G and cloud computing to applications such as games and virtual events (Huynh-The et al. 2023). At the infrastructure level, AI enhances technologies by improving network performance and security through resource allocation, traffic management, and fault detection. In terms of user interaction, AI powers smart input-output systems for devices like VR headsets and smart glasses, predicting user movements through sensors. It also supports realistic computer modeling, avatar rendering, and NPC animation. AI-driven machine vision helps XR devices recognize and analyze visual input, understand environments, and interpret user behavior.

AI also strengthens blockchain in the metaverse by improving security, scalability, and personalization (Jeon et al. 2022), and it supports the IoT through intelligent data processing. Additionally, AI enables natural language processing, allowing for voice recognition, real-time translation, and the creation of chatbots and virtual assistants that guide and engage users.

2.3.5 Blockchain

Blockchain technology was introduced in 2008 and was proposed soon thereafter as a key technology for building the metaverse. At its core, blockchain is a method of secure digital recordkeeping. In society, records—like receipts, marriage certificates, and identity documents—are essential for verifying events and claims, typically maintained by trusted third parties such as banks or governments. Blockchain offers an alternative by creating secure, verifiable records without relying on intermediaries. Using cryptography, decentralization, and consensus, it enables peer-to-peer transactions on the Internet. It underpins cryptocurrencies like Bitcoin and Ethereum, and is also used in smart contracts, real estate, personal data security, and supply chain tracking.

Blockchain is a type of database that works like a ledger or account book, storing records called blocks. Each block contains data about transactions or events—such as a Bitcoin transfer from one user to another—along with a timestamp and a link to the previous block using its unique identifier, or hash. This creates an ordered, growing chain of records. Blockchains are distributed across computer networks, with each node holding an identical copy, hence the term "distributed ledger technology." Most blockchains are public, allowing anyone to copy the ledger and add new blocks.

Blockchains ensure trust through cryptography and decentralization. Each block has a cryptographic hash based on its contents, so any tampering changes the hash—the unique identifier of a block —and breaks the chain. While this could theoretically be bypassed with enough computing power, blockchain's distributed nature adds another layer of security. Consensus algorithms help validate the blockchain by enabling network peers to collectively agree on its accuracy, without needing a central authority. This combination of cryptography and peer consensus secures the integrity of the data.

Blockchain is often seen as a key technology for the metaverse because it offers several valuable functions. First, it enables secure verification of ownership for digital assets and secure value transfer, mainly through smart contracts and cryptocurrencies. Second, it supports interoperability, allowing digital assets to move across platforms. Third, it aids in secure identity management for users and avatars. Fourth, blockchain can promote open standards and non-proprietary protocols, fostering transparent, collaborative metaverse development. Fifth, it enables decentralization, shifting control from single companies to users, making the metaverse more democratic. Experts have acknowledged that blockchain is not absolutely necessary for the development of the metaverse (Singer 2022). Nevertheless, it does offer many benefits and has been embraced by many developers and advocates as a key metaverse technology.

2.3.6 Internet of Things

The current Internet connects computers, but it can also connect everyday devices—like refrigerators, cars, or lamps—so they can exchange data with computers and each other. To do this, devices need *embedded systems*: small, dedicated computer systems, often in the form of chips, that control physical operations and handle data input and output. These systems receive data from the Internet and send information about the device's state back. They rely on *sensors* to detect internal or environmental conditions and convert them into signals, and *actuators* to carry out actions based on digital instructions. Sensors can also be connected to the Internet independent from any device, for instance, to measure temperature, ocean currents, or body signals and transmit that data online.

The IoT is the network of things that are connected to each other and to the Internet through embedded technology and sensors. It is already in widespread use, with billions of devices online. IoT data is processed by software, often without human involvement, to either control devices or collect data for analysis, planning, or decision-making. To interpret this data, IoT systems often use AI, especially machine learning, to generate insights, predictions, and instructions. Devices that combine IoT and AI are called "smart devices" and are capable of context-awareness, adaptation, and personalized responses.

The IoT is a key metaverse technology for several reasons. First, metaverse technologies like VR and AR rely on smart, connected devices—headsets, gloves, controllers—which require IoT to function effectively. Second, IoT data enhances virtual modeling and animation by supplying real-world information for precise and realistic 3D reconstruction.

Third, IoT enables real-time simulations, especially for digital twins. Digital twins require real-time sensor data from their physical equivalent, which the IoT can provide. The IoT can also provide real-world sensor data to shape things and events in virtual worlds, for example, by simulating realistic weather conditions in a game or by using measurements of a user's emotional state to influence the looks and behavior of her avatar. Fourth and finally, the IoT can be used to send instructions from the metaverse to things in the physical world. It may be used to generate virtual interfaces to simplify device control, or to monitor and control a production line via its digital twin, or to send data about a virtual design to a 3D printer.

2.3.7 Infrastructure Technologies

The metaverse demands powerful IT infrastructure to process massive amounts of data quickly and with minimal latency. This includes technologies like 5G/6G, cloud computing, advanced CPUs/GPUs, and new methods like edge and fog computing. Simulating responsive virtual worlds in real time requires constant, high-volume processing—far beyond current capabilities. Raja Koduri, a vice president of Intel, has stated that "what we imagined in *Snow Crash*, what we imagined in *Ready Player One*, for those experiences to be delivered, the computational infrastructure that is needed is 1000 times more than what we currently have" (Nover 2021). He is confident, however, that industry will be able to deliver on the needed infrastructure.

A major challenge is developing semiconductors capable of processing trillions of operations per millisecond—potentially one petaflop within 1–10 milliseconds (Nover 2021). Further miniaturization of chips, sensors, and wearables is also needed. On the network side, 5G, with speeds of up to 4 Gbit/s, may be sufficient to support the metaverse, though 6G is already in development (Alsharif et al. 2020). The metaverse will also demand more robust cloud computing, as hosting 3D environments requires immense processing and storage. Businesses may rely on specialized metaverse clouds with enhanced graphical capabilities (Tozzi 2022). Additionally, edge computing, considered vital for the metaverse (Carlini 2022), brings processing and storage closer to users through localized data centers, reducing latency and conserving bandwidth.

2.3.8 Conclusion

We reviewed seven key technologies essential to building the metaverse. Of these, VR and AR, along with 3D modeling and animation, are defining technologies. In essence, the metaverse consists of immersive, interconnected virtual worlds created through 3D design and accessed via VR or AR. These worlds interact with the physical world through AR and the IoT, operate on high-speed networks like 5G/6G, and are powered by cloud and edge computing with advanced hardware. Blockchain supports transactions in the metaverse, while AI is used to make everything work smarter and more efficient.

2.4 So What Is the Metaverse?

The concept of the metaverse that emerges from our analysis can be summed up as follows: The metaverse is a network of immersive, persistent, interconnected, interoperable, shared 3D virtual worlds that are, by preference, accessed through

VR systems, and may also include shared virtual objects and scenes that are accessed through AR. The metaverse will be interconnected with the physical world through AR and the IoT. The metaverse will consist of virtual worlds, and networks thereof, that are multifunctional as well as specialized. It will serve a large number of functions, serving as a platform for social interaction, entertainment, commerce, work, education, healthcare, and others. The metaverse will constitute a significant part of the Internet and, if VR and AR interfaces were to replace traditional, screen-based interfaces, it might even become the Internet.

The metaverse will likely rely on a mix of open and proprietary standards and be developed, owned, and operated by multiple parties. It remains unclear whether it will be dominated by large companies or take a more decentralized, user-driven form. Multiple, loosely connected metaverses may emerge—divided by ideology, language, intellectual property, or target audience—such as industrial, enterprise, and consumer-focused metaverses, or centralized vs. decentralized ones. Additionally, many VR/AR applications and virtual environments will exist outside the metaverse proper. These include VR and AR applications for individual and offline use, VR and AR environments that are not persistent but ad hoc and temporary, and non-immersive virtual worlds.

An ambiguity in current uses of the term "metaverse" is that the term is used to refer to a totality of networked immersive virtual worlds, but it is also used to refer to single immersive virtual worlds. So, the virtual worlds Second Life, Roblox, and Decentraland have been called metaverses, but the metaverse is also the envisioned future integration of these virtual worlds and others into one interoperable network. As long as interoperability has not been realized, single immersive virtual worlds may be referred to as metaverses, but once it has been established, the metaverse should be a name for integrated networks of immersive virtual worlds. Individual worlds in such networks should then simply be called virtual worlds.

The metaverse will be populated by avatars and non-player characters (NPCs), though some users may participate without avatars, using familiar tools like video calls or text chat. While MR headsets will be common, others will access it via traditional screens and keyboards. In professional settings, avatars will likely be photorealistic and resemble the user, while in playful contexts, they may be stylized or fictional. In AR, users that share a physical space will either be perceived as they appear in the physical world or as photorealistic avatars without visible headsets.

As defined here, the metaverse refers to shared, social virtual environments accessed via VR or AR. Some broaden the term to include any 3D-rendered environment, including single-user VR/AR, but this stretches the concept. A strong case can be made that shared, persistent, and social interaction is central to the metaverse. More open to debate is whether the concept should also apply to social virtual environments that are offline, that are nonpersistent or that are restricted to small groups of users. Some may want to stretch the concept this way, while others may opt for a more restrictive usage.

References

Alsharif, M. H., A. H. Kelechi, M. A. Albreem, S. A. Chaudhry, M. S. Zia, and S. Kim. 2020. Sixth Generation (6G) Wireless Networks: Vision, Research Activities, Challenges and Potential Solutions. *Symmetry* 12:676–697. https://doi.org/10.3390/sym12040676.

Avanade. 2022. Cutting Through the Metaverse Hype. https://www.avanade.com/-/media/asset/point-of-view/avanade-metaverse-pov.pdf. Accessed 27 Mar 2025.

Bartle, Richard. 2003. *Designing Virtual Worlds*. New Riders.

Bezmalinovic, T. 2022. The EU Commission Wants to Regulate the Metaverse. *Mixed Reality News*, September 19. https://mixed-news.com/en/the-eu-commission-wants-to-regulate-the-metaverse/. Accessed 27 Mar 2025.

Carlini, Steven. 2022. How Edge Computing Will Power the Metaverse. *Forbes*, May 18. https://www.forbes.com/sites/forbestechcouncil/2022/05/18/how-edge-computing-will-power-the-metaverse/. Accessed 28 Mar 2025.

Che, C. 2022. The Top 10 Metaverse Companies in China. *The China Project*, February 15. https://thechinaproject.com/2022/02/15/the-top-10-metaverse-companies-in-china/. Accessed 27 Mar 2025.

Clegg, N., and J. Olivan. 2021. Investing in European Talent to Help Build the Metaverse. *Meta*, October 17. https://about.fb.com/news/2021/10/creating-jobs-europe-metaverse/. Accessed 27 Mar 2025.

Coogan, C. G., and B. He. 2018. Brain-Computer Interface Control in a Virtual Reality Environment and Applications for the Internet of Things. *IEEE Access* 6 (10840–10): 849. https://doi.org/10.1109/ACCESS.2018.2809453.

Cureton, D. 2022. Siemens, NVIDIA Partner to Build Industrial Metaverse. *XR Today*, July 4. https://www.xrtoday.com/mixed-reality/siemens-nvidia-partner-to-build-industrial-metaverse/. Accessed 27 Mar 2025.

European Commission. 2022. People, Technologies & Infrastructure—Europe's Plan to Thrive in the Metaverse. September 14. https://ec.europa.eu/commission/presscorner/detail/en/STATEMENT_22_5525. Accessed 27 Mar 2025.

Fabernovel. 2022. Into the China Verse: How the Meta Wave Is Hitting China. https://asia.fabernovel.com/2022/05/13/into-the-chinaverse-how-the-meta-wave-is-hitting-china/. Accessed 27 Mar 2025.

Global Times. 2024. China Brings Together Tech Giants, Universities to Draft Metaverse Standards. January 21. https://www.globaltimes.cn/page/202401/1305805.shtml. Accessed 27 Mar 2025.

Google Trends. 2022. Google Trends for "metaverse" (Worldwide). https://trends.google.com/trends/explore?date=2021-10-01%202022-10-01&q=metaverse. Accessed 27 Mar 2025.

Huynh-The, T., Pham, Q., Pham, X., Nguyen, T., Han, Z. and Kim, D. 2023. Artificial intelligence for the metaverse: A survey. *Engineering Applications of Artificial Intelligence* 117:105581. https://doi.org/10.1016/j.engappai.2022.105581.

Interesse, Giulia. 2023. China Releases Three-Year Action Plan for Metaverse Industry Development. *China Briefing*, September 25. https://www.china-briefing.com/news/china-releases-three-year-action-plan-for-metaverse-industry-development/. Accessed 27 Mar 2025.

Jeon, H., Youn, H., Ko, S., and Kim, T. 2022. Blockchain and AI Meet in the Metaverse. *Advances in the Convergence of Blockchain and Artificial Intelligence* 73–82. https://doi.org/10.5772/intechopen.99114.

Kovach, Steve. 2022. Microsoft Is Selling the Metaverse Now — and It's Helping Make Everything from Robots to Ketchup. *CNBC*, May 24. https://www.cnbc.com/2022/05/24/microsoft-partners-with-kawasaki-for-industrial-metaverse.html. Accessed 27 Mar 2025.

Lowry, P. B., W. Boh, S. Petter, and J. M. Leimeister. 2024. Long Live the Metaverse: Identifying the Potential for Market Disruption and Future Research. *Journal of Management Information Systems* 62.

Metaverse Standards Forum. n.d. What Is the Mission of the Metaverse Standards Forum. https://metaverse-standards.org/faq/. Accessed 27 Mar 2025.

References

Milgram, Paul, and Kishino, Fumio. 1994. A taxonomy of mixed reality visual displays. *IEICE Transactionsons on Information and Systems* 77(12):1321–1329.

Nover, S. 2021. Intel Wants to Take You Inside the Metaverse. *Quartz*, December 13. https://qz.com/2101581/intel-is-ready-to-talk-about-the-metaverse/. Accessed 27 Mar 2025.

Parisi, T. 2021. The Seven Rules of the Metaverse. *Metaverses*, October 23. https://medium.com/meta-verses/the-seven-rules-of-the-metaverse-7d4e06fa864c. Accessed 27 Mar 2025.

Perri, Damiano, Marco Simonetti, Sergio Tasso, and Osvaldo Gervasi. 2023. Open Metaverse with Open Software. In *International Conference on Computational Science and Its Applications*, 583–596. Springer.

Sharma, Pardeep. 2024. Is the Metaverse Dead? What Happened to All the Hype? *Analytics Insight*, April 12. https://www.analyticsinsight.net/latest-news/is-metaverse-dead-what-happened-to-all-the-hype. Accessed 27 Mar 2025.

Sherman, W. R., and A. B. Craig. 2003. *Understanding Virtual Reality*. San Francisco, CA: Morgan Kaufmann.

Singer, A. 2022. Does the Metaverse Need Blockchain to Ensure Widespread Adoption? *Cointelegraph*, July 4. https://cointelegraph.com/news/does-the-metaverse-need-blockchain-to-ensure-widespread-adoption. Accessed 27 Mar 2025.

Smart, J., J. Cascio, J. Paffendorf, C. Bridges, J. Hummel, J. Hursthouse, and R. Moss. 2007. Pathways to the 3D Web: A Cross-Industry Public Foresight Project. https://www.w3.org/2008/WebVideo/Annotations/wiki/images/1/19/MetaverseRoadmapOverview.pdf. Accessed 27 Mar 2025.

Speicher, M., Hall, B. D., and Nebeling, M. 2019. What is mixed reality? *Proceedings of the 2019 CHI Conference on Human Factors in Computing Systems* 1–15. https://doi.org/10.1145/3290605.3300767.

Stephenson, N. 1992. *Snow Crash*. Bantam Books.

Tess. 2024. What Happened to the Metaverse? *Digivate*, June 7. https://www.digivate.com/blog/technology/what-happened-to-the-metaverse/. Accessed 27 Mar 2025.

Tozzi, Christopher. 2022. How the Metaverse May Impact Cloud Computing. *ITPro Today*, Apr 12. https://www.itprotoday.com/cloud-computing-and-edge-computing/how-metaverse-may-impact-cloud-computing. Accessed 28 Mar 2025.

World Economic Forum. 2022. *New Initiative to Build an Equitable, Interoperable and Safe Metaverse*. May 25. https://www.weforum.org/press/2022/05/new-initiative-to-build-an-equitable-interoperable-and-safe-metaverse/. Accessed 27 Mar 2025.

Zuckerberg, Mark. 2021a. Founder's Letter, 2021. *Meta*. https://about.fb.com/news/2021/10/founders-letter/. Accessed 27 Mar 2025.

Zuckerberg, Mark. 2021b. Meta (Facebook) connect 2021 metaverse event transcript. Interview by C. Sanford. Accessed 27 Mar, 2025. https://www.rev.com/blog/transcripts/meta-facebook-connect-2021-metaverse-event-transcript

Chapter 3
Will There Be a Metaverse?

> "We are at the beginning of the next chapter of the internet (…) Our hope is that within the next decade, the metaverse will reach a billion people, host hundreds of billions of dollars of digital commerce, and support jobs for millions of creators and developers."—Meta (2021)

Abstract This chapter of *The Metaverse: A Critical Assessment* addresses the question of whether—and under what conditions—a metaverse will emerge. It begins by examining the driving forces behind its development, focusing on potential benefits for users and society. These include general advantages, such as enhanced cognitive support and novel experiences, as well as sector-specific improvements in areas like healthcare, commerce, and education. The chapter argues that the metaverse offers meaningful benefits over current technologies.

It then considers the business case, concluding that while technological and investment challenges are likely surmountable, user acceptance may prove the most significant barrier. The analysis identifies key obstacles in user acceptance and explores how they might be overcome. Next, the chapter situates the metaverse within four broader technological and societal trends, each of which supports the plausibility of its eventual realization. These trends highlight the metaverse not as an isolated phenomenon but as part of a wider digital, social, and economic evolution. The chapter concludes with a tentative answer to the central question—"Will there be a metaverse?"—offering a balanced assessment of its prospects as a future digital platform.

Keywords Metaverse · Societal and user benefits · Application domains · Drivers and barriers to adoption · Business case · Technological viability · User adoption and acceptance

3.1 Introduction

In the previous chapter, we saw that serious efforts from the industry to develop the metaverse started in 2021. We also saw that there was a lot of hype surrounding the metaverse in 2021 and 2022, and that in 2023 and 2024, much of this hype subsided since the technology was not yet mature enough, practical applications fell short, and generative artificial intelligence stole the metaverse's thunder and became the new hype. Additionally, we saw that in spite of the shift in focus to artificial intelligence (AI), many tech companies still remained committed to the development of key metaverse technologies such as virtual reality (VR) and augmented reality (AR), and many industry insiders still retain a belief that the metaverse will emerge as a mass medium.

The future of the metaverse remains uncertain—its existence, timeline, and ultimate form are all still in question. If it does emerge, it is likely to evolve gradually, progressing from early proto-metaverses to highly advanced environments that closely mirror the physical world. I propose that its development could unfold in four key phases:

- *Phase 1. The early metaverse.* This is the phase we are currently in. This is a phase marked by the emergence of numerous VR-enabled virtual worlds and a growing number of digital twins in science, engineering, and healthcare. This period is characterized by extensive experimentation with both immersive and non-immersive environments beyond gaming—spanning social interaction, events, office meetings, collaboration, education, creative economies, virtual real estate, worldbuilding, e-commerce, and various hybrids. Smart glasses and AR/MR (mixed reality) headsets are seeing increased adoption, alongside a rising number of AR applications. While open standards are beginning to take shape, integration across virtual worlds remains in its early stages. The metaverse at this point is largely audiovisual (see our discussion of immersion in the previous chapter), with only limited use of haptic technologies.
- *Phase 2. The intermediate metaverse.* At this stage, developers have a clearer sense of viable metaverse business models. Technologies have matured and become more integrated, supported by major gains in processing and throughput speeds. VR and AR continue to gain acceptance, with new virtual worlds and AR applications built on established standards. The first interoperable virtual world networks have emerged, and VR/AR integration has significantly improved. Traditional interfaces like displays, keyboards, and mice may begin to be phased out in favor of VR and AR alternatives. Full-body tracking becomes standard, and the use of haptic technology is on the rise. Full-body tracking becomes the standard in the use of VR and AR, and the use of haptics is increasing.
- *Phase 3. The advanced metaverse.* Expansive, fully interoperable networks of virtual worlds now support a wide range of human activities. Some virtual worlds have grown so socially and economically complex that they function as societies, complete with institutions and democratic decision-making procedures. AR is widely used in workplaces, schools, and everyday life. The metaverse is now

fully haptic, with common use of haptic suits, treadmills, and motion simulators. Pan-sensory VR/AR is gaining popularity, and brain-computer interfaces are integrated alongside traditional inputs like hand, eye, and body movement. The line between VR and AR is fading—along with, for younger generations, the boundary between the virtual and the real.

- *Phase 4. The hyperreal metaverse.* Further development of the advanced metaverse culminates in the hyperreal metaverse. This is a metaverse in which a level of hyperrealism has been attained for all senses, making virtuality indistinguishable from reality. Very advanced VR/AR technology is used, possibly based on neural interfaces. The distinction between reality and virtuality has become largely inconsequential and is disappearing.

This is an ambitious scenario, in which the technology reaches full maturity and there is mass adoption. In an alternative scenario, metaverses will materialize at some point in the future, but they only have a limited user base and tend to be specialized and niche. Possibly, gaming metaverses are realized with tens or hundreds millions of users, but non-gaming applications, like office work, shopping, education, and healthcare, still largely take place outside of shared virtual environments, in the physical world and through conventional internet applications.

In this chapter, we explore both driving forces behind the emergence of the metaverse and the obstacles that may hinder its development. In the next section, we will discuss potential benefits that the metaverse could bring over conventional digital technologies and that could drive their adoption. In Sect. 3.3, we will discuss the business case for the metaverse, and possible barriers that may hinder its adoption. In Sect. 3.4, we will identify four broader trends within which the development of the metaverse may be positioned and which each point toward its eventual development. In Sect. 3.5, finally, we will arrive at some tentative answers to the question: "Will there be a metaverse?"

3.2 Why the Metaverse? Potential Benefits for Users and Society

What benefits could the metaverse bring to users and society? The metaverse consists of online platforms that either add to or displace current internet platforms. Its benefits could therefore best be understood as benefits it would have over the current internet (Web 2.0). The current internet already serves many functions that are also associated with the metaverse: information and communication functions, social networking, gaming and entertainment, creation and artistic expression, business meetings, and collaborative work, among others. The question that we will seek to answer is: which of these functions can be performed better by the metaverse and how? And are there any quantitatively new functions that the metaverse could perform? We will first provide a general account of the potential benefits of the metaverse for users and society, followed by an account of benefits for particular application domains.

3.2.1 General Benefits for Users and Society

One of the greatest benefits of the metaverse is undoubtedly that it provides *better possibilities for social interaction* than the current internet. This is especially true in a mature metaverse with realistic, full-body avatars and fully interactive virtual environments. Static avatars in low-interaction settings would not offer the same benefits. Unlike the current internet—which relies mostly on written or oral communication through tools like messaging, email, or video calls—the metaverse allows users to communicate with body language, facial expressions, and movement in shared 3D spaces. Users can walk, gesture, move closer to or farther from others, and engage in activities while conversing. Group dynamics are more fluid, allowing people to naturally form or shift conversations. Interaction is also enhanced by the presence of shared virtual objects and environments. For example, an architect could display a 3D model of a building or walk clients through a full-scale digital version, turning conversation into an interactive experience.

The metaverse also offers far better support for social activities than the current internet, which remains limited in this domain. While today's internet allows for online games, video calls, and information and resource sharing, it falls short in enabling shared, immersive experiences. The metaverse, by contrast, can host nearly any social activity—often replicating or even enhancing physical-world experiences. Events like family gatherings, parties, reunions, or playdates can take place in realistic 3D environments that match the occasion. Remote families can share a meal around a virtual table, high school classmates can reunite in a digital garden or ballroom, and friends can attend concerts, explore virtual destinations, or take part in cultural and creative activities together. Even group sports are possible with VR support like treadmills.

In short, the metaverse enables richer, more embodied social interaction than the current internet—both for communication and shared activities. In some cases, it may even surpass the physical world, offering enhancements like adjustable concert sound or seamless movement in crowded virtual spaces. But social interaction is not the only advantage. Let us now consider benefits the metaverse offers beyond co-presence.

A second benefit of the metaverse for users is that it can provide *better support for cognitive tasks*. Cognitive tasks are tasks that involve mental processes directed at acquiring knowledge and understanding, such as thinking, planning, calculating, perceiving, reading, categorizing, and remembering. Such understanding is gained through thoughts and our perceptions of, and interactions with, the environment. The metaverse can support these processes in ways that the current internet cannot. It can make information more easily accessible and easier to organize and process. Data can be visualized in 3D and easily interacted with using one's hands or eyes. Internet pages and files can be placed in 3D space and easily resized or removed with one's hands or the gaze of one's eyes. Moreover, the metaverse offers interactive and dynamic 3D models of things and places. These are more informative than pictures or text, and can be used for learning, planning, and decision-making.

Information storage can also be improved, by replacing the 2D desktop and its 2D folders and files with a 3D storage environment. These forms of support can benefit not only single persons but also groups who engage in joint cognitive tasks like brainstorming, planning, and analyzing (Kye et al. 2021).

A third benefit of the metaverse is that it can provide *better support for creative processes*, from fine arts to engineering design. The metaverse offers powerful means for visualization and for creating interactive models that are part of a creative process. These can be manipulated, tested out, and improved by the user, who can interact with them and relate to them with their hands and body. The immediate products of this are virtual rather than physical. However, virtual structures and processes can function as plans and blueprints for physical realization. Additionally, among other products, there are media products like movies, digital paintings, and computer games, which need no physical realization. Once again, these benefits can be realized for both individuals and collaborative teams.

Fourth, the metaverse is good at *delivering experiences*, more so than the current internet. The very idea of VR and AR is that of experiencing objects and environments as if they are really present, so this is what the metaverse is good at. The metaverse can provide all kinds of interactive experiences that have value to users because they are pleasant, beautiful, insightful, educational, imaginative, spiritual, or unique. These can include, for example, the experience of sitting at a beautiful beach, alone or with a friend, visiting bustling Times Square in New York and avoiding collisions with other pedestrians and cars, exploring the Louvre or Vatican museums but without the crowds, traveling to Mars in a spaceship and setting the first foot on the planet, or taking on the form of an elephant and living and interacting with other elephants in a herd. Some of these experiences will be solitary, some will be shared. Some will be passive, while others will require active engagement from the user. These experiences can occur in the context of a game, an educational program, a social gathering, or any other activity in the metaverse.

Fifth, the metaverse can support *better ways of interacting with the physical world*. One way it can do so is through the Internet of Things (IoT), which connects virtual and augmented reality to physical reality through sensors (devices that detect qualities in the physical environment) and actuators (devices for motion or control). The IoT can support digital twins, which are real-time virtual simulations of physical systems and processes, and the information gained from digital twins can enable more successful engagement with their physical counterpart, for instance, by signaling the need for repairs. The metaverse can also present users with virtual interfaces, in either virtual or augmented reality, that connect to actuators in the physical world, in order to better interact with objects and systems in the physical world. For example, a virtual interface in VR can be used to issue commands to a robot in the physical world. This process is called teleoperation. Lastly, AR can also contain virtual enhancements of, and information about, objects and systems in the physical world, thereby allowing for more successful interaction with them. For example, in surgery, blood vessels can be highlighted in real time, and live information about the patient's condition can be projected in the surgeon's visual field.

Sixth and finally, a blockchain-enabled metaverse can support users by *enabling peer-to-peer transactions that do not require a trusted third party*. This cuts costs and makes users less dependent on other parties for their transactions. It potentially also enhances security, privacy, and transparency.

It is mainly through these six types of benefits, for both individuals and groups, that the metaverse will benefit society at large. They translate to benefits for organizations in both the private and public sector and to benefits for social institutions, including healthcare, education, science, and the economy. Next, we will discuss the benefits of the metaverse within key social institutions and economic sectors, in a section on application domains.

3.2.2 Benefits within Specific Application Domains

We will now consider potential benefits of the metaverse in seven major application domains. They are media and entertainment, education, commerce, office work, manufacturing, healthcare and transportation.

3.2.2.1 Media and Entertainment

The benefits of metaverse technology are already quite apparent in media and entertainment, where social interaction and immersive experiences converge. The metaverse offers vast possibilities for entertainment (Shazhaev 2022)—from next-level gaming and immersive films to concerts with spectacular effects and up-close views of performers. Virtual theme parks could go beyond physical ones, taking users on thrilling rides through oceans, ancient cities, or along volcanoes. AR can also enhance entertainment in the physical world, adding effects to people, buildings, and events.

As for media, the metaverse will add immersive, interactive content to its repertoire. Mass media could deliver interactive, immersive stories in 3D, even with tactile feedback. As for social media, the metaverse can expand on text-based, image-based, and video-based social media with a 3D, immersive, and interactive format (Filipchuk 2022). Moreover, since the metaverse supports communication formats that resemble face-to-face communication, the distinction between social media and regular social interaction may blur. For instance, avatars of influencers may appear in the immediate vicinity of users as if users are actually meeting with them.

3.2.2.2 Education

Education is another major application of the metaverse, which builds on its strengths in social interaction, cognition, creativity, and immersive experience. Compared to the current internet, the metaverse can offer virtual classrooms that are far more interactive and engaging than video calls or static online content, allowing for richer student-teacher and peer interactions (Kye et al. 2021). Moreover, intermittent use of VR and AR applications may also provide significant benefits in physical classroom settings by providing interactive simulations for learning (Elmqaddem 2019). In history class, for example, the metaverse may immerse pupils in simulations of Alexandria in the third century BC, or nineteenth century London, and may even have them play roles to give them an idea of what it was like to live in these times. In biology class, the metaverse may allow AR-enhanced physical specimens. In physics class, AR may be used to visualize complex data sets, and in engineering class, students may use an AR or VR to explore digital twins. These kinds of interactive simulations, if realistic and accurate, enable experiential learning, an important form of learning which involves doing things and then reflecting on the experience.

3.2.2.3 Commerce

Commerce is often hyped as the application domain of the metaverse that is potentially the most profitable. The metaverse could become a major space for buying both physical and virtual goods and services. Unlike current e-commerce, which lacks the ability to inspect or try products, the metaverse allows users to engage with life-sized, realistic virtual replicas (Behzad and Jain 2022). Shoppers can try on virtual clothing or preview furniture in their homes, reducing returns and improving satisfaction. The metaverse also enables more social, personalized shopping experiences with interactions between customers and human or AI assistants. Business-to-business commerce could benefit similarly through interactive product simulations and digital twins.

Many service industries could also operate in the metaverse, especially those in media, education, training, consulting, and finance. Some services requiring physical interaction—like restaurants, gyms, or salons—may not immediately work online, but advanced VR with haptics and sensory feedback may eventually simulate aspects of these experiences as well. AR, meanwhile, can enhance real-world services with digital overlays and interactive features.

The metaverse also supports a growing market for virtual goods—such as digital real estate, furniture, art, and fashion—as users and businesses spend more time in immersive environments. Past virtual worlds have shown that robust virtual economies can emerge around such goods.

The metaverse also offers new opportunities for advertising, marketing, and branding. With users more immersed, companies can potentially collect more information about them and their preferences (de Ruyter et al. 2020). Companies can also disseminate virtual samples of products to gauge customer demand before manufacturing them at scale. And finally, they can explore new ways of advertising and branding. Next to displaying banners and billboards in the metaverse, they can also project 3D models and animations, stage virtual events, engage in product placement in virtual experiences, sell virtual wearables with their brand, and issue virtual collectibles. With AR, companies can project billboards and banners into the physical world, highlight their products when they come into view, and provide additional information about them.

3.2.2.4 Office Work

Office work is another major application domain for the metaverse. The metaverse can provide several benefits compared to the current internet. First, it can provide a better support for meetings and for collaborative work. Indeed, as we already argued in our earlier discussion of benefits, the metaverse can provide better support for meetings than video conferencing and other communication tools. In addition, meeting participants can use immersive 3D models for joint planning, analysis, and brainstorming. In data-intensive professions, the metaverse can support interactive data visualization in 2D and 3D, making communication and decision-making more effective. The metaverse can also provide good support for virtual conferences and social team meetings. And lastly, as we saw in our discussion of education, the metaverse can also support the training of employees, in part through interactive simulations of tasks.

3.2.2.5 Manufacturing

For the manufacturing industry, proposals have been made to develop a separate industrial metaverse (Kannan 2022). This metaverse is focused on the faithful simulation of products, systems, and processes for the purposes of design and development, testing, monitoring, and maintenance. Industrial metaverse environments would allow for the visualization of 3D designs that enable the creation or modification of products, production facilities, and production processes in a fast, easy, and cost-effective way. For products and facilities that are already in existence, digital twins can be developed, with real-time data coming in from the physical counterpart, to test and monitor products and systems, explore possible redesigns, and predict malfunction and the need for maintenance. AR also has applications in manufacturing, to enhance work in production facilities through enhanced visualization of systems and tools, by providing real-time information as well as virtual interfaces. These technologies can also be used for training purposes.

3.2.2.6 Healthcare

The potential benefits of the metaverse for healthcare are also all about visualisation, interactive simulation, and digital twins. A first application area is the use of AR in surgery, where AR could improve surgical outcomes by superimposing imagery and information in the surgeon's view. The use of VR and AR in training can have similar benefits (Yang et al. 2022). Immersive, interactive radiology imaging can provide doctors with better information, as can 3D data visualisation. As a last benefit, the metaverse can enable new forms of digital therapeutics, using VR and AR for the evaluation and treatment of psychiatric disorders, support groups, and even—with the addition of haptics—for physical therapy. For example, VR therapy can be used to treat acrophobia and agoraphobia by exposing patients to virtual heights and populated areas, and can potentially also treat gender dysphoria by means of different avatars and roleplay in virtual environments that allow for gender experimentation (Sears 2020; Davis 2022). VR has also proven to be effective in improving coordination and balance in physical therapy.

3.2.2.7 Transportation

The metaverse could reduce the need for transporting both people and goods. People often travel to meet others, access amenities, or visit places—all of which the metaverse can simulate with high fidelity. Instead of spending time and money on travel, users may choose to interact, work, or attend events in virtual environments. Likewise, if people and businesses increasingly engage in virtual settings, like virtual offices, theme parks, and concerts, demand for physical goods—and their transport—could decline. This shift could hurt the transportation sector but benefit individuals, businesses, and the environment through time, cost, and emissions savings. It is worth noting that these reductions may not actually materialize, though. The internet was also expected to reduce transportation at one point, but these reductions did not materialize. As for environmental benefits, the metaverse will require a lot of energy, potentially offsetting some of the gains made through reductions in transportation.

3.3 The Business Case for the Metaverse and Potential Barriers

While the benefits outlined in the previous section suggest potential for a metaverse business case, perceived advantages alone are not enough to justify one. A compelling business case requires a clear problem statement, technical and operational feasibility, financial viability, consideration of risks and barriers, and a viable path to implementation. In this section, we will consider the business case for the metaverse, first considering the expected economic impact of the metaverse, and considering potential barriers and whether and how they might be overcome.

3.3.1 Potential Barriers

Studies have estimated that by 2030, the global gross economic impact of the metaverse could be anywhere from 1.1 trillion to as high as 13 trillion US dollars.[1] These are big numbers, especially when taking into account that by 2030 the metaverse will likely still be in its early stages. Among the largest markets expected by analysts are e-commerce, manufacturing, healthcare, education, and gaming (Elmasry et al. 2022; PwC 2019). The numbers quoted by these studies could, of course, be far from accurate. Nevertheless, what our earlier review of benefits and applications has found is that there could be significant demand. Demand ultimately comes from consumers and business customers, as metaverse end-users. They are the ones who pay for the devices, software, and services that are needed for metaverse access and they are the ones who subsequently purchase, rent, or lease digital and physical goods and services in the metaverse. The judgment of these studies is that, provided the pace of technological innovation and investment continues as it has been until the early 2020s, the supply of the technology and services will be there, and the demand will be sufficient to support continued investment.

Elmasry et al. (2022) distinguish four categories of industry that will benefit from the metaverse. Their first category is *infrastructure and hardware*. This covers hardware and software infrastructure and devices for running the metaverse, such as networks, cloud, VR headsets, accessories, and operating systems that are part of the human interface. (See also the technology section, and particularly the category "infrastructure" in Sect. 3.2 of Chap. 3.) Industries in this category would benefit from the metaverse, as long as there is sufficient overall demand for metaverse technologies. A second category is *platforms*, which include tools and platforms for building 3D content, and platforms facilitating the distribution and discovery of content, such as metaverse search engines and app stores. A third category is *enablers*, which are platforms and services that support security, privacy, identity management, governance, payments, and monetization. Demand in these two categories will generally also follow overall demand for the metaverse.

The fourth and final category is *content & experiences*, which is where much of the action will be. This includes virtual worlds, applications, and other types of content, such as creator content and user-generated content. Applications, in this study, are software programs tied to specific use cases such as education, collaborative work, and industrial design. In this category, the success of industry players will be determined less by general demand for metaverse services, and more by demand for the specific product that they provide. A virtual world or application that supports services for which demand turns out to be limited may not be successful.

---

[1] Christensen and Robinson (2022), an often cited study, put the global gross economic impact at 3 trillion US dollars by 2031 and a share in global GDP of 2.8%. Fortune Business Insights (2022) places it at 1.5 trillion by 2029, Elmasry et al. (2022) at 4 to 5 trillion by 2030, and Ghose et al. (2022) even at 8 to 13 trillion by 2030. A more recent report, Global Industry Analysts (2025), only puts the impact by 2030 at 1.1 trillion.

What appears to be missing from this taxonomy are regular businesses that are customers of the content and experiences industry, and that engage in commerce in the metaverse or use it to support business operations. These include companies that open virtual shops, use virtual offices, sponsor events, rely on third-party digital twins, or offer virtual services—all of which carry direct or indirect economic value. Some may also develop their own applications, blurring the line between content providers and business customers. For example, the home furnishings company IKEA has released an AR app that lets users project its furniture into their homes, effectively acting as a content provider.

What barriers might prevent the metaverse from emerging or succeeding? We now turn to these challenges and their potential impact. A key concern is technological limitations: can a fully functional metaverse be built with current or near-future technology? Many developers believe the necessary technologies are already available or close to maturity. This is true for areas including 3D modeling and animation, blockchain, AI, and the IoT (see also Chap. 3). As for virtual and augmented reality, the technology has been rapidly improving in recent years. The Apple Vision Pro (2024) and the Meta Quest 3 (2023), the two leading headsets in the mid-2020s, exemplify high-performance VR/MR through high-resolution display quality, spatial computing capabilities, and responsive user interaction.

A key limitation of current headsets is that they only support what was called the audiovisual metaverse in the previous chapter: they provide visual and auditory input but lack full-body tracking and haptic feedback through gloves or suits. As a result, users experience only partial embodiment, with limited movement and no sensation of object resistance—reducing the sense of presence. VR/AR researcher Louis Rosenberg calls this the problem of perceptual inconsistency: for example, trying to lean on a virtual table that is not physically there causes a mismatch between virtual and real-world sensations (Smith 2022). While full-body tracking and haptic devices exist and can improve immersion, it is still unclear whether they can fully overcome this challenge. AR is less affected, as it enhances the physical world rather than replacing it.

The biggest current technological barrier may be infrastructure. A seamless metaverse will require advanced networks (5G, 6G), expanded cloud and edge computing, and next-generation CPUs and GPUs—demands that will take years and major investment to fulfill. Surveys show that network quality is especially critical for users, who expect low-latency, uninterrupted virtual experiences (Amdocs 2022). While 5G is generally seen as sufficient for metaverse bandwidth, only 25% of global mobile connections were 5G by the end of 2024 (Ericsson 2024), and just 25% of the global population is expected to have access to ultra-fast high-band 5G by 2030 (Grijpink et al. 2020). The rollout of new computing infrastructure will also take time. While infrastructure poses a short-term hurdle, it is unlikely to be a long-term obstacle, as improvements in connectivity, processing, and storage have historically advanced at a steady pace.

Another technological limitation could be the absence of interoperability standards that enable the interconnection of virtual words and support the evolution of the industry. While some of such standards already exist, most still have to be developed, and it is uncertain that parties will be able to arrive at strong agreements. However, it is promising, in this respect, that the newly formed Metaverse Standards Forum, discussed earlier in this chapter, is supported by almost the entire industry.

A second potential barrier is *lack of investor readiness*. Investors need to believe in the business case for the metaverse, in its technological feasibility, and in the presence of sufficient user demand. Otherwise, the needed investments will not be made. In the early 2022, the signs were that investors are ready and willing. Elmasry (2022) reported US$57 billion of investments in the metaverse in 2021, followed by $120 billion in the first five months of 2022 alone. These investments decreased in 2023 and 2024, as enthusiasm for the metaverse waned and the generative AI hype set in. Nevertheless, reports by leading consultancy firms in 2024 and 2025 still project a large market for the metaverse in the coming decade. In addition, the technologies needed for the metaverse have use cases separate from it and are drawing large investments regardless of confidence in the metaverse.

A third potential barrier is *lack of adoption by users (or user acceptance)*. Studies indicate that there is a significant willingness to try out the metaverse, and a belief that metaverse technologies will shape the future. Surveys show that over half of U.S. consumers would consider replacing in-person experiences with metaverse alternatives if connectivity allows (Amdocs 2022), and 50% of the global population views VR and AR positively (Ipsos 2022). The same study shows that majorities of the global population believe that the metaverse will transform education, entertainment, work, healthcare, and socialization.

However, curiosity and a belief in potential benefits do not always translate into sustained use. Potential users could still reject the technology for other reasons than a perceived lack of benefits. Technological products may be rejected because the costs are deemed too high, because there are perceived downsides and risks, and because there is a perceived lack of quality and operational readiness. As for costs, these could be a concern, as one study shows that 26% of US consumers would not be willing to pay more than $100 for metaverse hardware (Amdocs 2022). Historically, though, mass production of IT hardware and software has been able to keep costs down so that computers and internet access are affordable for the majority of consumers. Likewise, quality and operational readiness may be initial concerns that will likely be alleviated over time, as the technology improves.

Even if cost is not a barrier for most users, a range of risks and downsides could still hinder metaverse adoption. Safety and security are key concerns: in a study of perceived metaverse risks, a majority of U.S. adults expresses worry about issues such as data tracking and misuse, online abuse, cyberbullying, personal safety, sexual harassment, and motion sickness in VR. Data privacy stands out, with 55% citing it as a major concern (Clement 2022). These findings suggest that unless safety is clearly addressed, many potential metaverse users may remain hesitant. Another notable concern is addiction to simulated realities, with 47% of global internet users expressing anxiety about this risk (Statista Research Department 2022). Note, though, that many of these risks also exist on today's internet, yet they have not stopped widespread adoption. This suggests users may make similar trade-offs with the metaverse, accepting certain risks in exchange for perceived benefits.

The most important barrier to the metaverse may lie in the fact that many adult users do not seem to enjoy spending large amounts of time in VR. Second Life pioneer Philip Rosedale has said that his experiences over many years are that adults do not like to inhabit animated avatars and spend a lot of time in virtual worlds (Wilde 2022). Similarly, VR and AR pioneer Louis Rosenberger has argued that people do not like to spend a lot of time in VR, and he therefore expects that the VR metaverse will draw a relatively small audience (Anderson and Rainie 2022). Part of the issue is that people do not like to be closed off from physical reality for an extended period of time. Another issue is that people experience discomfort and fatigue with extended use of a headset.

Could these barriers barrier be overcome? In part, they could be the result of conservatism of older generations that will disappear over time as younger generations take their place. Studies show that younger generations are much more interested in VR than older ones, and much more likely to use it. One study showed that while only 6% of baby boomers and only 18% of Gen-X report VR usage, 38% of millennials and 45% of Gen-Z do (Boland 2022). As the technology matures, moreover, aversions against using it may subside. Aversions against the use of avatars could result from the fact that current versions are cartoonish and not fully embodied, and they may lessen with the advent of photorealistic and fully embodied avatars. Possibly, adults still associate virtual environments with gaming and entertainment and have not been convinced yet of their value in professional work, but this could change as better applications are developed. Possibly, also, future VR and AR systems will be more comfortable to wear, and the advent of MR will keep users connected to their physical environment, if they so please, while they engage with the metaverse, making for a better experience. At this point in time, though, there is no denying that a significant barrier to the metaverse is found in usability, particularly in relation to ergonomics and perceived usefulness.

3.4 Four Trends Toward the Metaverse

We now turn to four major trends that provide context for the emergence of the metaverse. These are ongoing technological and societal developments in which the metaverse can be seen not only as a natural progression, but in some cases a culmination. The trends include the evolution toward a next-generation internet (Web 3.0 or Web3), the shift toward natural human-computer interaction, the rise of interactive, shared virtual worlds, and the broader societal movement toward the geographical disembedding of goods and services. We will examine each in turn, highlighting how the metaverse aligns with and extends these developments.

3.4.1 The Metaverse and Web 3.0

The first trend is the ongoing effort to conceptualize and build the next-generation internet. It is widely recognized that the Internet has evolved through two major phases—Web 1.0 and Web 2.0 (Choudhury 2014; Investopedia 2022). Web 1.0, spanning roughly from 1989 to 2004, is often described as the "read-only" web. During this phase, websites were prepared by webmasters, they were static, and users were passive consumers of information. Around 2004, the rise of social media marked the transition to Web 2.0—the "read-write" or "read-write-execute" web—characterized by dynamic, user-generated content, collaboration, personalization, and interactivity. This phase emphasized participation over presentation and was further propelled by the rise of the mobile Internet, which expanded access and usability across devices.

Since the emergence of Web 2.0, people have been speculating about the next incarnation of the Web, Web 3.0. For a long time, Web 3.0 has been equated with World Wide Web inventor Tim Berners-Lee's 1999 concept of a Semantic Web. The Semantic Web was conceived of as an advanced Internet controlled by software agents, AI programs that are able to intelligently retrieve and process information for users. The term "Web 3.0" was coined in 2006 by reporter John Markoff to refer to the Semantic Web. While this vision of a future Internet still has many adherents and finds some validation due to the rise of generative AI, several other visions of the future Internet have since emerged. We will discuss three of the most salient ones, all of which have immediate relevance to our understanding of the metaverse.

A second vision has long been rather marginal but has recently picked up steam. This is a vision of a future Internet based on three-dimensional (3D) graphics and interactions, going beyond the two-dimensional content of Web 2.0: the "3D internet" or "3D Web". The effort to deliver 3D content on the Web has existed since at least the mid 1990s. One of its main advocates, the Web3D Consortium, was formed in 1997 with the mission to support interactive 3D content embedded into web pages (Web3D). The consortium played a key role in the introduction of an international technical standard for Web3D content in 1997, VRML, and its 2004 successor X3D. The gaming industry has also provided a boost to 3D Web content, with the first online 3D games—Doom and Quake—being introduced in the mid-1990s.

However, these efforts rarely resulted in claims that 3D should be considered the future of the Internet; more so that it would be an interesting add-on. A more radical vision was presented in 2007 within the already mentioned Metaverse Roadmap Project, whose subject was "Pathways to the 3D Web". Its key 2007 roadmap document stated: "In time, many of the Internet activities we now associate with the 2D Web will migrate to the 3D spaces of the metaverse." (Smart et al. 2007). An academic publication in the same year also made a case for this evolution (Alpcan et al. 2007). Much later, in 2019, an influential book by Gabriël René and Dan Mapes promoted the notion of the Spatial Web (René and Mapes 2019). The Spatial Web was defined as a future information infrastructure that was to involve many recent (metaverse) technologies, with 3D modeling, as well as VR and AR, at the core of its vision of the future Internet. The book also referred to this future Internet as "Web 3.0".

A third vision of the next-generation Internet is the IoT. Since the term was coined in 1999, innovators have envisioned an internet that extends beyond computers to include everyday objects connected through embedded sensors and communication technologies. The goal has been to enable smarter monitoring and control of the physical environment. Kreps and Kimppa (2015) argue that the IoT represents a form of Web 3.0, given its radical expansion of what the Internet encompasses. Large-scale integration of devices began in the 2000s, and by 2008–2009, more "things" were connected to the Internet than people (Evans 2011). In many ways, the IoT is already a reality—though often operating in the background of daily life. To the extent that IoT defines a post–Web 2.0 Internet, it has already substantially arrived.

Finally, the fourth and currently influential vision of a future Internet is that of a decentralized, blockchain-enabled web, also called Web3 (and sometimes Web 3.0). The notion of Web3 was introduced by Polkadot founder and Ethereum co-founder Gavin Wood in 2014 (Edelman 2021). Wood observed that all major services in Web2 (also known as Web 2.0), including those involving user-generated content, are in the hands of a few tech companies that exert centralized ownership and control over them. He envisioned a future Web, enabled by public blockchain, which did not require such centralization, but would instead support decentralization and collective ownership: Web3, or the "read-write-own web". This vision became very influential by the early 2020s.

It should be noted that these four visions of the Internet point to an emerging view of the next-generation Internet as being 3D and immersive, blockchain-enabled, driven by AI, and powered by the IoT. All four of these visions refer to technologies that are considered central to the metaverse: 3D modeling, virtual and augmented reality, blockchain, IoT and AI. In this way, the metaverse can be considered as the convergence of some key visions of the future Internet that have emerged since the beginning of the twenty-first century. Collectively, these visions support a conception of the future Internet in which the metaverse is, if not constitutive, at least a large part of it.

3.4.2 The Metaverse and Natural Interaction

A second trend is the trend toward more natural, embodied, and immersive human-computer interaction. What is natural interaction? Let us consider how people normally interact with their environment. People are immersed in environments that contain other people, physical objects, and structures. They interact with them through bodily engagement, including motion, gesture, and speech. Computing, however, initially required interactions that were largely disembodied and nonimmersive. Operating computers in the 1950s, 1960s, and 1970s was a considerable mental exercise of thinking up symbolically encoded instructions, inserting them into the computer through a keyboard or other input device, and awaiting a response on a monitor or other output device.

Over the years, there has been a trend toward human-computer action that is progressively more natural, embodied, and immersive. The advent of graphical user interfaces in the 1980s, with the Macintosh and Microsoft Windows operating systems for home computers, and the arrival of mice as input devices, made human-computer interaction more natural by making interactions more similar to interacting with the physical world. The transition toward multimedia computers around the same time, with image, sound, and video capability, further enhanced this effect. The arrival of touchscreen interfaces in the 2000s, thanks to which graphical objects could be operated with one's fingers rather than with a keyboard or mouse, was another step toward more natural, embodied interfaces. So was the inclusion of sensors in mobile devices, such as GPS sensors, gyroscopes and accelerometers, which allowed for computing that was more aware of the physical environment and the position and movements of the body.

The increasing use of 3D graphics in the 1990s and beyond resulted in interactive environments that had a still greater similarity to the physical world. Their engagement from a first-person perspective through avatars, in video games and open 3D worlds, further enhanced a sense of natural, embodied interaction. Over the years, increased realism in 3D modeling and animation has added to the feeling of immersion and embodiment in such worlds, and the introduction of (massively) multi-user options has also added the possibility of semi-embodied interaction (through avatars) with other human beings. Finally, VR enables fully immersive environments that are potentially interacted with using the full body, including all the senses. At its best, VR does not present its users with an interface. Instead, its input and output devices are made invisible to the user, retreated into the body and the virtual environment, and interaction is just like interaction with the physical world. With the addition of a rich, fully interactive, multi-user world, which the metaverse is intended to be, natural human-computer interaction is potentially realized to its fullest. In short, it can be concluded that the metaverse can be understood as the culmination of a trend toward more natural, embodied, and immersive human-computer interaction.

3.4.3 The Metaverse and Online Worldbuilding

A third trend relevant to the metaverse is the growing use of computers and the internet for simulation—especially the creation of *virtual worlds*, which are computer-generated environment where users interact with each other and digital surroundings through *avatars*—digital representations of users. Since their inception, computers have been used for simulation (Bartle 2010). For instance, during World War II, they were used to model the process of nuclear detonation within the scope of the Manhattan project. Early simulations were scientific and engineering models designed to predict and explain the behavior of physical systems. These were typically non-interactive—the user was not part of the simulation—and lacked graphical representations of the systems being modeled.

This changed in the 1970s with the rise of computer graphics and video games, which introduced interactive simulations aimed at everyday users. Gaming became the primary driver of these developments, evolving into today's virtual worlds (Bartle 2010, 2016; Donovan 2010). Notably, *arcade games* and later *home consoles* like Pong, Space Invaders, Pac-Man, and Super Mario Bros brought immersive, graphical environments to the mainstream, allowing users to interact with virtual spaces through game figures (like a person, animal, automobile or spaceship), or else through a cursor or virtual tool.

In addition, multi-user games called *MUDs* (Multi-User Dungeon) presented users with simulated interactive environments in which they could encounter and interact with other users. Most MUDs present users with a fantasy world in which they play a character. More so than video games, MUDs can be considered precursors of the metaverse, due to their simulation of full-blown interactive environments that are also populated by other users.

The first MUD, called *MUD1*, was released in 1978 as a text-based game with no graphics. Users interacted with the environment through typed commands like "go north" or "pick up stick," receiving textual descriptions in response. MUDs remained text-only until the mid-1980s, when graphical elements from video games were combined with MUDs' multi-user features, resulting in early graphical MUDs such as *Habitat* (1987) and *Neverwinter Nights* (1991). These made use of avatars—graphical representations of users—as a means of interacting with virtual worlds. Initially 2D, graphical MUDs moved into 3D with games like *Meridian 59* (1996), often cited as the first 3D MUD. By then, the Internet had become a mass medium, enabling hundreds of users to play simultaneously. Around this time, the term "MUD" began to give way to "RPG" (Role-Playing Game), and Internet-enabled MUDs became known as *MMORPGs* (Massively Multiplayer Online Role-Playing Games), often featuring persistent worlds that continue evolving even when players are offline.

In the mid-1990s, multiplayer options for online play also started being added for computer games other than MMORPGs, like Doom and Quake, which are first-person shooter games. Some of these non-RPG online games only allowed a handful of players, while others also allowed hundreds of players simultaneously. These developments blurred the distinction between MMORPGs and other online multiplayer games and gave rise to the category of *MMOs*: massively multiplayer online games. MMOs include MMORPGs, but also other online games that allow many players at once but that may not involve role-playing or avatars.

By the mid-1990s, some MMORPGs began to develop social dimensions that extended beyond gameplay. They became platforms for interpersonal and group relationships, with players forming organized groups—such as clans, factions, or guilds—that often resulted in real friendships. Some games simulated entire towns or villages, enabling interaction with both other players and computer-controlled non-player characters, mimicking aspects of real-world social structures. Players could also own homes, craft items, build structures, and participate in emergent virtual economies based on the exchange of in-game goods. *World of Warcraft* (2004) was among the first to integrate many of these features and reached over 12 million players at its peak. A more recent example is *Fortnite* (2017), which has attracted over 400 million registered users.

MMORPGs like World of Warcraft and battle royale games like Fortnite have many of the features that can be expected in a metaverse: they are 3D virtual environments, they have large numbers of users that interface with each other through avatars, they are persistent, and they support social and economic activity. However, they are still essentially games, in that their main activity is gameplay, involving challenges, goals, competition, rules governing play, and outcomes that count as losing or winning.

Further steps toward the metaverse emerged through open-world and sandbox games. *Open-world games* offer non-linear gameplay, emphasizing exploration and player-driven objectives, as seen in titles like *Grand Theft Auto III* (2001), the *Fallout* series (1997–), and *The Elder Scrolls* series (1994–). *Sandbox games* also emphasize non-linearity but focus more on creativity and open-ended interaction, often without specific goals. Examples include *The Sims* (2000), which simulates domestic life, and *Minecraft* (2011), where players explore and reshape a virtual world by mining, building, and crafting. These two genres often overlap, combining exploration, creativity, and freedom. Still, both remain games in the minimal sense—structured activities involving challenges, objectives, or quests, primarily designed for entertainment.

In 2003, an open-world, multi-user, persistent 3D virtual world was released that did not define itself as a game. This was *Second Life*. Created by Linden Lab, it offered a space where users—called "residents"—could create avatars, explore a vast virtual world of continents and cities, socialize, participate in events, build property, and engage in commerce. Unlike traditional MMOs, *Second Life* emphasized open-ended living rather than structured gameplay. It hosted everything from art exhibitions and concerts to educational programs, religious services, and business meetings. The platform also reflected real-world complexities, including the occurrence of virtual theft, gambling, virtual prostitution, assault and riots. Its economy was driven by the Linden Dollar, which could be exchanged for real currency. At its peak, *Second Life* had over one million active users and is regarded by some as the first true metaverse—despite lacking immersive VR, which many consider a defining feature.

By the time the metaverse concept gained traction in 2021–2022, MMO gaming was already widespread, with many titles boasting tens of millions of users. However, most lacked the fully open-ended, sandboxed nature and social-economic depth of *Second Life*. Only a small number more closely resembled the metaverse vision, emphasizing open-world design, user-generated content, and social interaction. While most of these had modest user bases—typically in the hundreds of thousands or low millions—some, especially those targeting younger audiences, attracted much larger communities. Many of these virtual worlds integrated blockchain, cryptocurrencies, and tradeable digital assets such as virtual buildings, avatar wearables, and digital artwork.

Let us consider some of the most popular metaverse-like platforms of the early to mid-2020s. One of the most prominent is *Roblox*, launched in 2006, which enables users—primarily children under the age of 12—to create and explore user-generated 3D virtual worlds. Players appear as avatars and can participate in games, virtual events, and social gatherings, with interaction supported through both text and voice chat. Roblox features a virtual economy and regularly hosts community-driven activities such as parties and playdates. In 2020, it was reported that over half of American children were using the platform (Lyles 2020), and its growth surged during the COVID-19 pandemic, as it provided a safe online space for social interaction. By 2024, Roblox had 380 million monthly active users and 80 million daily active users (Backlinko 2025).

Habbo (2000; formerly Habbo Hotel) is a series of virtual worlds aimed at teens and young adults. In it, users meet as avatars in public or private rooms of a large hotel to chat and role-play. They can also build and design rooms, take care of virtual pets, create and play games, craft items, and buy and sell items. In 2022, Habbo had over 300 million registered users and close to a million active monthly users.

Decentraland (launched in 2020) is a virtual world built on blockchain technology, offering a degree of decentralized ownership and governance. Users can buy virtual land through its marketplace, with ownership registered as NFTs on the Ethereum blockchain. These landowners can build structures, host events, trade digital assets, and vote on platform development. Some collaborate on shared neighborhood projects, while major brands like Samsung, Adidas, and Miller Lite have also acquired land. Decentraland has hosted events such as virtual fashion weeks—featuring brands like Dolce & Gabbana and Tommy Hilfiger—and concerts by well-known artists. In 2021, it had over 300,000 monthly active users, primarily adults, but by 2024, this had dropped to the low tens of thousands. Other blockchain-based virtual worlds with similar features include *The Sandbox* (2012), *Voxels* (2018), *Axie Infinity* (2018), *Bloktopia* (2022) and *Otherside* (2023).

While many of these platforms have metaverse-like qualities, they are often not considered "true" metaverses because they are not primarily experienced through immersive VR. However, some—such as Decentraland and Roblox—do support VR headsets. Other platforms have been built specifically with VR in mind. *VRChat* (launched in 2014) is a leading social VR platform offering environments for socializing, exploration, worldbuilding, events, and commerce. In early 2025, it has around 900,000 daily users, making it one of the largest in its category (ActivePlayer 2025). *Horizon Worlds*, Meta's VR social platform, reached nearly 200,000 monthly active users by the same time (Batchelor 2022). *Sansar*, a VR-focused successor to *Second Life* developed by Linden Lab, was sold in 2020 after the company concluded that VR technology was not yet mature enough to support its vision (Summers 2020). Another notable platform is *HiberWorld* (launched 2020), a browser-based worldbuilding platform with over three million monthly active users, which added VR support in 2024.

Overall, these virtual worlds are primarily geared toward socializing and entertainment, offering limited support for professional work. Yet since the rise of the Internet, professional collaboration has increasingly moved online. Early tools were mostly asynchronous and text-based, but the 2000s and 2010s saw the steady growth of platforms enabling real-time communication and collaboration—such as video conferencing, 2D virtual event platforms, and 2D virtual offices. These gained widespread use during the 2020 pandemic, but their limitations also became clear: they are non-immersive, often disembodied, and can lead to fatigue when used extensively ("Zoom fatigue").

In response, the 2010s saw the emergence of virtual world–based alternatives aimed at work and collaboration. These platforms typically allow users to meet as avatars in shared 3D environments, and often have VR/AR support. Examples include *Mibo* (2020), designed for informal team meetings with playful interaction; *Horizon Workrooms* (launched in beta in 2022), Meta's VR-based virtual office that simulates meeting rooms; and *Microsoft Mesh* (2022), a MR extension of Microsoft Teams that enables avatar-based collaboration with persistent 3D content. Other enterprise platforms like *Virbela* (2018), *Spatial* (2018), *ENGAGE LINK* (2022) and *Arthur* (2022), offer immersive spaces for work, education, exhibitions, and events, often with support for both VR and AR.

The broader trend emerging from these developments is clear: since the rise of MUDs and video games in the 1970s, virtual worlds have steadily evolved toward increasingly metaverse-like forms. Key milestones include the shift from text-based to graphical and then 3D MUDs, the rise of massively multiplayer online games with the advent of the Internet, the adoption of open-world and sandbox designs, the integration of social and economic systems, the emergence of decentralized blockchain-enabled platforms, the creation of professional virtual environments, and the addition of virtual and augmented reality. By the early 2020s, some virtual worlds had accumulated so many of these features that it has become reasonable to debate whether they qualify as true metaverses.

The gaming industry has played a central role in driving this evolution. As Microsoft CEO Satya Nadella noted, game technologies are likely to be foundational to the metaverse (Waters 2022). Yet metaverse-like worlds are not solely rooted in gaming. Their social dimensions borrow from social media and online networking; their economic systems draw on e-commerce and blockchain innovation; and their professional capabilities build on tools from video conferencing, virtual events, and digital workspaces. In this sense, the metaverse emerges not just from gaming, but from the convergence of multiple Internet-based technologies and services, each with its own history and trajectory.

3.4.4 The Metaverse and Geographical Disembedding

The three trends toward the metaverse discussed so far reflect the evolution of digital technologies. However, there are also broader social and economic developments that help contextualize the metaverse. One of the most significant is the historical process of *geographical disembedding* (Brey 1998). This refers to the increasing separation of goods and services from their physical locations, allowing people to access them regardless of geography. It is the process by which amenities—goods, services, persons, or places that people want to have access to, experience, interact with, or use—can be made available to us with increasing ease. Amenities are often located elsewhere, creating barriers to availability. Over time, technological progress has reduced these barriers, by making remote access easier, safer, and more efficient, or by reproducing amenities or some of their features. The geographical disembedding thesis suggests that as technologies advance, they progressively enable access to once-distant amenities with less time, effort, and physical movement.

In Brey (1998), I identified several historical processes that contribute to geographical disembedding. The first is *time-space convergence*—the reduction of travel time and effort between locations due to advances in transportation and communication technologies. Innovations such as the automobile and airplane have shortened physical travel times, while the postal system, telegraph, and telephone have accelerated information exchange and communication, making distant amenities more accessible.

The second process, *space blending*, is an intensified form of time-space compression in which access to other locations becomes immediate. Technologies like the telephone, radio, and television enable real-time perceptual and communicative access across distances, effectively blending separate spaces. Beginning with the telephone (1876) and radio (1893), this process has accelerated with the rise of electronic media. More recently, the IoT and telerobotics have introduced remote physical interaction, further dissolving spatial boundaries.

The third stage is *space generation*: the creation of entirely new, computer-generated environments that transcend physical space. These virtual spaces—also called cyberspace or digital space—include websites, chatrooms, game worlds, and digital platforms that people can "visit" and interact with, even though they lack physical coordinates. These spaces offer place-independent access to amenities such as information, communication, services, and entertainment. In doing so, space generation pushes geographical disembedding even further than time-space convergence or space blending, radically increasing the availability of amenities without the constraints of physical location.

How does the metaverse fit into this historical process of geographical disembedding? Well, the current Internet provides its users with various amenities, but it does so with limitations. It gives access to them through computer displays that contain mostly text and 2D images and streams, and it lets users interact with them through cumbersome input devices like keyboards and mice. The metaverse promises better access to, and interaction with, amenities through 3D visualisation as well as VR and AR. Thus, instead of reading about a product and seeing a picture of it in an online store, one can visit a 3D webstore in VR and inspect a 3D rendering of the product before ordering it. Instead of communicating with one's friends through text or 2D video, one can meet them in VR, experience their whole presence in 3D, and even "touch" them with data gloves. In this way, the metaverse promises better availability of more amenities than the current internet and can in this way be understood as a next step in the process of geographical disembedding.

3.4.5 Other Trends and Conclusion

These are some of the main trends within which the emergence of the metaverse can be situated, but there are others. One is the trend toward more integration of the digital and physical world. Traditionally, activity on computers and the Internet has been largely disconnected from users' physical environments. The rise of the IoT represents a shift toward bridging this gap by enabling digital systems to interact directly with the physical world. Similarly, AR and MR technologies bring digital content into the user's immediate surroundings. The metaverse aligns with this trend, as it increasingly incorporates IoT and AR/MR to create more seamless interactions between the digital (virtual) and physical.

Another key trend involves the commercial motivations driving metaverse development. The metaverse represents a major business opportunity not just for tech companies but also for consumer brands and service providers. Sociologist Jan van Dijk argues that since the 1970s, we have transitioned from an industrial society to an information society. In this shift, businesses moved away from mass communication—ineffective for increasingly individualized consumers—toward highly segmented and personalized marketing. The Internet enabled detailed profiling and targeted advertising; the metaverse could take this even further. Because user behavior in virtual environments is tracked in granular detail, companies may gain unprecedented access to behavioral data, allowing for even more refined and personalized marketing strategies (Van Dijk 1993).

In conclusion, several influential trends can be pointed to explain the current interest in the metaverse. These include trends in the evolution of computing and the Internet, as well as broader social, economic, and technological trends. For each of these trends, it was shown how the metaverse can be seen as the logical next step.

3.5 Conclusion

The central question of this chapter was: "Will there be a metaverse?" After examining its drivers, barriers, and broader historical and technological trends, we are now in a position to offer a preliminary answer. First, we found that a fully realized metaverse offers clear advantages over today's internet and digital media. It promises enhanced social interaction, improved support for cognitive and creative tasks, more immersive experiences, deeper integration with the physical world, and more secure forms of digital exchange. The range of use cases is wide, extending beyond gaming and entertainment to include work, commerce, education, and healthcare.

We also observed that the metaverse appears technologically feasible and that early-stage platforms—or proto-metaverses—already exhibit many of its key features. There is significant interest among developers and investors, and hardware costs are likely to decrease over time, reducing price as a barrier for most users. Finally, we identified four major trends that provide a larger context for the metaverse's emergence: the shift toward Web 3.0, the move toward more natural human-computer interaction, the increasing realism and interactivity of virtual worlds, and the ongoing geographical disembedding of goods and services. Taken together, these findings suggest that the metaverse is not just a speculative vision, but a likely outcome of ongoing technological and societal evolution.

At present, the primary barrier to the emergence of the metaverse appears to be the limited usability and perceived usefulness of current VR and virtual world platforms. Many users find VR headsets uncomfortable for extended use, are reluctant to navigate environments through avatars for applications other than gaming and remain unconvinced of the value of these technologies beyond gaming and entertainment. These reservations may stem from the immaturity of the technology—a phase that could pass as VR becomes more immersive, intuitive, and aligned with compelling real-world use cases. Time will tell whether these concerns fade with progress. It is also important to note that younger users, in particular, tend to be more open and enthusiastic about these experiences.

Given the metaverse's vast potential and the wide applicability of its underlying technologies, continued development is highly likely. We can expect increasingly sophisticated metaverse-like platforms to emerge, extending well beyond gaming and entertainment into fields such as education, work, healthcare, and commerce. What remains uncertain, however, is the timeline for the technology's maturation and the scale of its eventual user adoption.

References

ActivePlayer. 2025. VRChat Live Player Count. https://activeplayer.io/steam/vrchat. Accessed 29 Mar 2025.
Alpcan, T., C. Bauckhage, and E. Kotsovinos. 2007. Towards 3D Internet: Why, What, and How? In *2007 International Conference on Cyberworlds*, 95–99. https://doi.org/10.1109/CW.2007.62.
Amdocs. 2022. What Do Consumers Want from the Metaverse? *Amdocs*. https://www.amdocs.com/sites/default/files/2022-06/Metaverse_Consumer_Research_FINAL.pdf. Accessed 28 Mar 2025.
Anderson, Janna, and Lee Rainie. 2022. *The Metaverse in 2040*. Pew Research Centre. https://www.pewresearch.org/internet/wp-content/uploads/sites/9/2022/06/PI_2022.06.30_Metaverse-Predictions_FINAL.pdf. Accessed 28 Mar 2025.
Backlinko. 2025. *Roblox User and Growth Stats You Need to Know*. https://backlinko.com/roblox-users. Accessed 29 Mar 2025.
Bartle, Richard A. 2010. From MUDs to MMORPGs: The History of Virtual Worlds. In *International Handbook of Internet Research*, ed. Jeremy Hunsinger, Lisbeth Klastrup, and Matthew Allen, 23–39. London, New York: Springer.
Bartle, Richard A. 2016. *MMOs from the Inside Out: The History, Design, Fun, and Art of Massively-Multiplayer Online Role-Playing Games*. Berkeley, CA: Apress.
Batchelor, James. 2022. Meta's flagship metaverse Horizon Worlds struggling to attract and retain users. *GamesIndustry.Biz*, October 17. https://www.gamesindustry.biz/metas-flagship-metaverse-horizon-worlds-struggling-to-attract-and-retain-users. Accessed 29 Mar 2025.
Behzad, S., and A. Jain. 2022. Metaverse—The Next E-commerce Revolution. *Deutsche Bank*, July 29. https://flow.db.com/more/technology/metaverse-the-next-e-commerce-revolution?language_id=1. Accessed 28 Mar 2025.

Boland, Mike. 2022. Who's Using VR, and How Often? *AR Insider*, October 5. https://arinsider.co/2022/10/05/whos-using-vr-and-how-often/. Accessed 28 Mar 2025.

Brey, Philip. 1998. Space-Shaping Technologies and the Geographical Disembedding of Place. In *Philosophy and Geography III: Philosophies of Place*, ed. Andrew Light and Jonathan Smith, 239–263. New York, London: Rowman & Littlefield.

Choudhury, N. 2014. World Wide Web and Its Journey from Web 1.0 to Web 4.0. *International Journal of Computer Science and Information Technologies* 5 (6): 8096–8100.

Christensen, L., and A. Robinson. 2022. *The Potential Global Economic Impact of the Metaverse*. Analysis Group. https://www.analysisgroup.com/Insights/publishing/the-potential-global-economic-impact-of-the-metaverse/. Accessed 28 Mar 2025.

Clement, J. 2022. U.S. Metaverse Concerns 2022. *Statista*, April 25. https://www.statista.com/statistics/1303391/us-adults-potential-metaverse-issues/. Accessed 28 Mar 2025.

Davis, Nicola. 2022. VR Role-Play Therapy Helps People with Agoraphobia, Finds Study. *The Guardian*, April 6. https://www.theguardian.com/technology/2022/apr/05/vr-role-play-therapy-helps-people-with-agoraphobia-finds-study. Accessed 28 Mar 2025.

De Ruyter, K., J. Heller, T. Hilken, M. Chylinski, D. I. Keeling, and D. Mahr. 2020. Seeing with the Customer's Eye: Exploring the Challenges and Opportunities of AR Advertising. *Journal of Advertising* 49 (2): 109–124. https://doi.org/10.1080/00913367.2020.1740123.

Donovan, Tristan. 2010. *Replay: The History of Video Games*. East Sussex, England: Yellow Ant.

Edelman, Gilad. 2021. The Father of Web3 Wants You to Trust Less. *Wired*, November 29. https://www.wired.com/story/web3-gavin-wood-interview/. Accessed 29 Mar 2025.

Elmasry, T., H. Khan, L. Yee, E. Hazan, G. Kelly, R. Zemmel, and S. Srivastava. 2022. *Value Creation in the Metaverse*. McKinsey & Company. https://www.mckinsey.com/~/media/mckinsey/business%20functions/marketing%20and%20sales/our%20insights/value%20creation%20in%20the%20metaverse/Value-creation-in-the-metaverse.pdf. Accessed 28 Mar 2025.

Elmqaddem, N. 2019. Augmented Reality and Virtual Reality in Education. Myth or Reality? *International Journal of Emerging Technologies in Learning* 14 (3): 234–242. https://doi.org/10.3991/ijet.v14i03.9289.

Ericsson. 2024. Mobile Subscriptions Outlook. https://www.ericsson.com/en/reports-and-papers/mobility-report/dataforecasts/mobile-subscriptions-outlook. Accessed 28 Mar 2025.

Evans, Dave. 2011. *The Internet of Things: How the Next Evolution of the Internet Is Changing Everything*. Cisco Internet Business Solutions Group. https://www.cisco.com/c/dam/en_us/about/ac79/docs/innov/IoT_IBSG_0411FINAL.pdf. Accessed 29 Mar 2025.

Filipchuk, Yurii. 2022. The Metaverse as the Next Step in Social Media Evolution. *LinkedIn*, August 17. https://www.linkedin.com/pulse/metaverse-next-step-social-media-evolution-yurii-filipchuk-/?trk=public_post. Accessed 28 Mar 2025.

Fortune Business Insights. 2022. Metaverse Market Size & Share Forecast [2022–2029]. https://www.fortunebusinessinsights.com/metaverse-market-106574. Accessed 28 Mar 2025.

Ghose, R., N. Surendran, S. Bantanidis, K. Master, R. Shah, and P. Singhvi. 2022. Metaverse and Money: Decrypting the Future. *Citi*. https://ir.citi.com/gps/x5%2BFQJT3BoHXVu9MsqVRoMdiws3RhL4yhF6Fr8us8oHaOe1W9smOy1%2B8aaAgT3SPuQVtwC5B2%2Fc%3D. Accessed 28 Mar 2025.

Global Industry Analysts, Inc. 2025. *Metaverse – Global Strategic Business Report*. Research and Markets. https://www.researchandmarkets.com/reports/5548465/metaverse-global-strategic-business-report

Grijpink, F., E. Kutcher, A. Ménard, S. Ramaswamy, D. Schiavotto, J. Manyika, M. Chui, R. Hamill, and E. Okan. 2020. *Connected World: An Evolution in Connectivity Beyond the 5G Revolution*. McKinsey & Company. https://www.mckinsey.com/industries/technology-media-and-telecommunications/our-insights/connected-world-an-evolution-in-connectivity-beyond-the-5g-revolution. Accessed 28 Mar 2025.

Investopedia. 2022. Web 3.0 Explained, Plus the History of Web 1.0 and 2.0. October 23. https://www.investopedia.com/web-20-web-30-5208698. Accessed 28 Mar 2025.

References

Ipsos. 2022. *How the World Sees the Metaverse and Extended Reality.* Ipsos. https://www.ipsos.com/sites/default/files/ct/news/documents/2022-05/Global%20Advisor%20-%20WEF%20-%20Metaverse%20-%20May%202022%20-%20Graphic%20Report.pdf. Accessed 28 Mar 2025.

Kannan, Ramya. 2022. How the Metaverse Will Influence the Manufacturing Sector. *UST*, July 29. https://www.ust.com/en/insights/how-the-metaverse-will-influence-the-manufacturing-sector. Accessed 28 Mar 2025.

Kreps, D., and K. Kimppa. 2015. Theorising Web 3.0: ICTs in a Changing Society. *Information Technology and People* 28 (4): 726–741. https://doi.org/10.1108/ITP-09-2015-0223.

Kye, B., N. Han, E. Kim, Y. Park, and S. Jo. 2021. Educational Applications of Metaverse: Possibilities and Limitations. *Journal of Educational Evaluation for Health Professions* 18. https://doi.org/10.3352/jeehp.2021.18.32.

Lyles, Taylor. 2020. Over Half of US Kids Are Playing Roblox, and It's About to Host Fortnite-esque Virtual Parties Too. *The Verge*, July 22. https://www.theverge.com/2020/7/21/21333431/roblox-over-half-of-us-kids-playing-virtual-parties-fortnite. Accessed 29 Mar 2025.

Meta. (2021, October 28). *Founder's Letter.* https://about.fb.com/news/2021/10/founders-letter/

PwC. 2019. Seeing Is Believing: How Virtual Reality and Augmented Reality Are Transforming Business and the Economy. https://www.pwc.com/gx/en/technology/publications/assets/how-virtual-reality-and-augmented-reality.pdf. Accessed 28 Mar 2025.

René, Gabriel, and Dan Mapes. 2019. *The Spatial Web: How Web 3.0 Will Connect Humans, Machines, and AI to Transform the World.* Pub. Gabriel René and Dan Mapes.

Sears, Brett. 2020. Virtual Reality in Physical Therapy. *Verywell Health*, December 20. https://www.verywellhealth.com/vr-headsets-in-physical-therapy-and-rehab-5084948. Accessed 28 Mar 2025.

Shazhaev, Ilman. 2022. Metaverse: The Next Stop for Social Networking and Influencers? *Finextra Research*, March 7. https://www.finextra.com/blogposting/21945/metaverse-the-next-stop-for-social-networking-and-influencers. Accessed 28 Mar 2025.

Smart, J., J. Cascio, J. Paffendorf, C. Bridges, J. Hummel, J. Hursthouse, and R. Moss. 2007. Pathways to the 3D Web: A Cross-Industry Public Foresight Project. https://www.w3.org/2008/WebVideo/Annotations/wiki/images/1/19/MetaverseRoadmapOverview.pdf. Accessed 27 Mar 2025.

Smith, Matthew S. 2022. Is the Metaverse Even Feasible? *IEEE Spectrum*, March 21. https://spectrum.ieee.org/is-the-metaverse-even-feasible. Accessed 28 Mar 2025.

Statista Research Department. 2022. Dangers of the Metaverse According to Internet Users Worldwide in 2021.Statista, July 7. https://www.statista.com/statistics/1288822/metaverse-dangers/. Accessed 28 Mar 2025.

Summers, Nick. 2020. Why "Second Life" Developer Linden Lab Gave up on Its VR Spin-Off. *Engadget*, March 27. https://www.engadget.com/2020-03-27-why-second-life-linden-lab-sold-sansar.html. Accessed 29 Mar 2025.

Van Dijk, J. A. G. M. 1993. Communication Networks and Modernization. *Communication Research* 20 (3): 384–407. https://doi.org/10.1177/009365093020003003.

Waters, Richard. 2022. Microsoft Chief Hails $75bn Activision Deal as Grand Step Into Metaverse. *Financial Times*, February 3. https://www.ft.com/content/95e17671-a6ac-4dcc-bae7-e9f34af1349c?segmentid=acee4131-99c2-09d3-a635-873e61754ec6. Accessed 29 Mar 2025.

Wilde, Tyler. 2022. The Creator of Second Life Has a Lot to Say About All These New 'Metaverses'. *PC Gamer*, April 2. https://www.pcgamer.com/second-life-metaverse-interview/. Accessed 28 Mar 2025.

Yang, D., Zhou, J., Chen, R., Song, Y., Song, Z., Zhang, X., Wang, Q., Wang, K., Zhou, C., Sun, J., Zhang, L., Bai, L., Wang, Y., Wang, X., Lu, Y., Xin, H., Powell, C. A., Thüemmler, C., Chavannes, N. H., Chen, W., Wu, L., Bai, C. 2022. Expert consensus on the metaverse in medicine. *Clinical EHealth* 5: 1–9. https://doi.org/10.1016/j.ceh.2022.02.001.

Chapter 4
Will the Metaverse Respect Our Rights?

> *"In countries like Canada and the US, we are guaranteed certain rights by law. For example: the right to own property, the right to privacy and the freedom to transact and engage in commerce in a lawful manner. Why shouldn't those rights extend online, and into virtual spaces?"[Tapscott (2024)]—*
> *Alex Tapscott, author and entrepreneur*

Abstract A major concern surrounding the metaverse is that users may face significant threats to their individual rights, including pervasive surveillance, manipulative immersive advertising, and various forms of harassment or abuse within virtual environments. This chapter of *The Metaverse: A Critical Assessment* explores these risks and considers strategies for protecting five key rights: security, privacy, freedom, equality, and property. In immersive environments, *security* extends beyond traditional cybersecurity to include harassment and stalking that can feel immediate and personal. *Freedom* encompasses not only speech but also movement, behavior, and presence—each vulnerable to monitoring, censorship, or manipulation. *Privacy* is under intense pressure, as platforms can collect and analyze biometric, behavioral, and emotional data in real time. *Equality* is at risk due to barriers to access to virtual spaces, embedded biases in platform design, and biases in moderation practices, potentially reinforcing real-world disparities. Finally, *property rights* grow more complex as users invest in virtual assets and identities that may lack clear ownership protections and remain subject to platform control. The chapter highlights the need for proactive safeguards to ensure that individual rights are upheld as immersive digital environments continue to evolve.

Keywords Metaverse · Human rights · Risks · Security · Privacy · Freedom · Equality · Property

4.1 Introduction

As discussed in the previous chapter, many prospective users have concerns about the risks the metaverse may pose. These include, among others, potential invasions of privacy through biometric data collection, virtual harassment by other users, addiction to metaverse use, manipulation through hyper-targeted advertising, and financial losses from scams or stolen digital assets. Such risks can be broadly divided into two categories: risks to individual rights and risks to well-being. This chapter will focus on risks related to rights, while the next will examine risks to physical, psychological, emotional, and social well-being. That chapter will also address broader societal risks, including potential threats to social and political institutions and the environment. Although the emphasis in both chapters is on identifying and mitigating harms, potential benefits of the metaverse will also be noted—albeit more briefly, as these were explored in depth in Chap. 3.

Rights are entitlements that people have which allow them to act in certain ways or be treated in certain ways by others. Rights are important because they ensure access to essential resources, opportunities and protections, and help to foster a stable and functional society. The concern in this chapter is specifically with human rights, universal entitlements that apply to all people by virtue of being human. Human rights include civil and political rights, which are specific types of human rights that protect personal freedoms and democratic participation. Rights are enshrined in legal systems and constitutional frameworks, and in international declarations and treaties, such as the *Universal Declaration of Human Rights* the *International Covenant on Civil and Political Rights* and are often also embedded in social norms and customary practices.

In this chapter, we examine five fundamental human rights that are at risk in the metaverse: the rights to security, privacy, freedom, equality, and property. These rights are widely recognized as essential—particularly in democratic societies—and are often enshrined in national legal systems. We will examine the potential threats to these rights in the metaverse and consider strategies for mitigating them. Many of these challenges are likely to mirror those encountered on the current internet and in the physical world. What the metaverse shares with the internet is that both are networked digital media that serve a wide range of human functions. What it shares with the physical world is the experience of embodied presence in a three-dimensional environment. Therefore, drawing parallels with similar issues in both digital and real-world contexts can offer valuable insights into the risks the metaverse poses to these fundamental rights.

4.2 Security in the Metaverse

Security is the protection of individuals, organizations, countries, and assets from crime, violence, or other harm caused by individuals and organizations. The right to security is essential because it protects the systems and environments in which other rights, like privacy, property, and freedom, operate. Without security, those rights can be easily undermined or violated. The security of persons is a fundamental individual right recognized in the Universal Declaration of Human Rights. It affirms that individuals have the right to bodily integrity and to protection from intentional physical or psychological harm inflicted by others. The security of assets is likewise safeguarded through legal protections, particularly property rights, which ensure individuals can control and defend their possessions.

In the field of digital technology, security threats have traditionally focused on assets—specifically, threats to information systems, networks, and data. These are known as cybersecurity threats. *Cybersecurity* refers to the practice of protecting digital networks, devices, and information from criminal or unauthorized access and use. This includes defense against well-known threats such as viruses, Trojan horses, denial-of-service attacks, and ransomware. Security on the internet is also largely focused on cybersecurity.

When comparing security issues in the metaverse with those on the broader internet, two key differences emerge. First, cybersecurity in the metaverse assumes new forms due to increased technological complexity. The use of richer and more diverse datasets, additional layers of technology, and more advanced input-output devices introduces novel vulnerabilities that go beyond traditional internet-based cybersecurity challenges. Second, the metaverse introduces new kinds of security risks that cannot be fully addressed under the conventional framework of cybersecurity. For example, users in the metaverse may be stopped, pushed, assaulted, or have their virtual possessions stolen. These actions do not compromise networks, systems, or data in the traditional sense, but instead threaten the security of persons and their virtual property. This highlights that security in the metaverse extends beyond protecting technological infrastructure—it must also account for interpersonal harm and violations within the immersive environment.

We refer to security threats arising from user actions within virtual worlds as *threats to virtual security*, and when these actions are criminal, we categorize them as *virtual crimes*. Virtual crimes take place within the virtual environment and typically target other users or their virtual assets. In contrast, *cybercrimes* generally occur outside the virtual world and involve attacks on underlying systems, software, or data, compromising cybersecurity. The metaverse, therefore, gives rise to two distinct but sometimes overlapping categories of crime: *virtual crimes*, which are internal to the environment and compromise the safety and autonomy of users and their digital property, and *cybercrimes*, which target the foundational technologies that support the metaverse. These categories can intersect—for instance, when a user manipulates others into disclosing sensitive information that is subsequently used in a cyberattack.

4.2.1 Cybersecurity Threats in the Metaverse

Many traditional cybersecurity threats that exist on the internet—such as spyware, viruses, Trojan horses, ransomware, denial-of-service attacks, social engineering, and unauthorized data breaches—are expected to persist in the metaverse. However, due to its increased technological complexity and access to new types of data, these threats may become more severe. The metaverse introduces not only new variants of existing threats but also entirely novel vulnerabilities that demand closer examination.

Data theft, to start with, becomes more concerning due to the generation of new types of data within the metaverse. Biometric information, virtual location data, and detailed records of user interactions can be highly privacy-sensitive and valuable for exploitation. Attackers who gain access to such data may use it to compromise identities, steal intellectual property (IP), or manipulate users. Additionally, hackers who control metaverse environments may alter virtual worlds in disruptive or harmful ways—causing avatars to change form, distorting physical laws, deleting entire environments, or using bots and haptic feedback devices to harass users physically and psychologically.

Social engineering also evolves in the metaverse. Social engineering is the manipulation of internet users to scam them and have them give up something valuable, like access codes, personal information, or valuable goods such as money and non-fungible tokens (NFTs). Unlike traditional approaches that rely on emails, texts or phone calls, attackers can deploy avatars or bots to establish deceptive relationships with users. These bots might impersonate friends or trusted figures by copying their avatars and behavioral patterns. Sentiment analysis tools can enhance these attacks by analyzing users' voice, gestures, or expressions to better tailor manipulation. Fake digital storefronts and maliciously coded virtual goods may further facilitate scams, embedding malware or extracting sensitive information.

Identity theft takes on new forms. Hackers may hijack or replicate avatars to deceive others, gaining access to resources, spaces, or relationships under false pretenses. Such impersonation can facilitate fraud, emotional manipulation, and even espionage, making it a particularly dangerous frontier in metaverse security.

The integration of the metaverse with the (Internet of Things) IoT also introduces new *cyber-physical threats*—security breaches in computer systems or networks that result in impacts on the physical environment. Devices like virtual reality (VR) headsets and haptic suits can be hacked to deliver unwanted sensory stimuli, causing discomfort or harm. Augmented reality (AR) systems could be manipulated to project misleading information into a user's physical surroundings. Digital twins—virtual replicas of real-world objects—pose additional risk, as attacks on them can affect the actual systems they represent.

4.2.2 Threats to Virtual Security in the Metaverse

We now turn to threats to virtual security, which was defined as security threats arising from user actions within virtual worlds that target people or property in a virtual world. These are harmful and often criminal acts committed by users of virtual worlds through their avatars. We will start with a discussion of criminal and harmful acts against persons that can be perpetrated in the metaverse. Let us first acknowledge that not all harmful and criminal acts against persons that are possible in the

4.2 Security in the Metaverse

physical world are possible in the metaverse. Any act that can only be defined in terms of physical acts and physical impacts cannot exist in the metaverse, because as a digital medium, it cannot deliver significant physical impacts on persons.

As a consequence, murder, manslaughter, and aggravated assault are not possible crimes in the metaverse. It is simply not possible for a user to carry out acts in the metaverse that directly bring about death or severe bodily injury. At best, a user can use the metaverse to manipulate or misinform other users in a way that indirectly leads to death or injury. Simple assault, however, assault that does not result in serious injury, might be a possible crime in the metaverse. This might include aggressive avatar behavior, simulated physical violence, and touching, grabbing, pushing, or hitting without consent, especially if the victim is making use of haptic devices. Likewise, intimidation, involving acts of making someone feel afraid or threatened, can also take place in the metaverse.

A similar distinction can be made for sex offenses. Many sex offenses appear to be possible in the metaverse. Some of them are exhibitionism (involving public sexual acts by avatars), sharing of sexually explicit content without consent, voyeurism (of sexual activity of users in private virtual environments), and grooming of minors. Sexual harassment, which involves repeated unwelcome sexual comments, gestures, or advances, is also possible in the metaverse.

More controversial are sexual assault and rape. These are acts that involve touching or penetrating the body of another person against their will. However, it may be objected, touching a user's avatar is not the same as touching their body. An avatar, it could be argued, is a representation or digital extension of a body, and not the body itself. Therefore, virtual sexual assault, which is directed at avatars, can at best be a form of sexual harassment, and not real sexual assault. However, this view downplays the fact that virtual sexual assault can feel like real sexual assault to the victim, due to the intimate, immersive relationship between the user and their avatar. This is especially the case when haptics are used, which allow both the perpetrator and the victim to not only hear and see, but also feel the other person. While legal systems may not recognize virtual sexual assault as sexual assault, its psychological and emotional impact may be similar.

It would seem that virtual rape cannot be real rape, since it involves, by most definitions, nonconsensual physical penetration, which cannot take place in digital space. Virtual rape is now often recognized as a form of sexual harassment. However, the use of teledildonics, which are networked, electronic sex toys to mimic human sexual interaction, could blur the distinction between simulation and reality. If a user makes use of an electronic sex toy, and a perpetrator takes unwanted control over such a toy, through hacking or by impersonating a trusted sex partner, then it could be argued that the resulting penetration is a form of rape.

Human trafficking is another serious crime that could manifest in the metaverse, albeit in digitally adapted forms. Traditionally, human trafficking involves modern-day slavery in which individuals are recruited, transported, or held through force, coercion, or fraud for the purpose of exploitation—most commonly in forced labor or commercial sex work—primarily for the economic benefit of traffickers. Such recruitment and forced employment could also take place in the metaverse. Victims could be put to work as virtual sex workers, using avatars to provide sexualized services, as virtual partners, friends, or companions in roleplaying or companion services, or be forced to create digital goods like clothing, objects, or NFTs for resale—similar to sweatshop labor, but focused on virtual economies.

A final category consists of harassment and bullying, which on the internet are mostly text- and media-based. In the metaverse, these behaviors shift to embodied interactions, making them more immediate and personal. This includes unwanted touching, sexual gestures, stalking, as well as the threat of such acts, delivered through avatars in real-time. Perpetrators can also harass users by commenting on their avatar's appearance or mimicking, mocking, or surrounding them in ways that feel invasive or abusive. The immersive nature of the metaverse makes these actions more intense and harder to ignore than traditional online harassment.

Next, we consider crimes against property in the metaverse, committed by users through their avatars. These crimes typically target the digital assets of individuals or organizations, including both traditional digital goods (like software, media files, and digital money) and virtual assets, which are 3D simulated entities such as virtual furniture, clothing, avatars, land, and buildings that can be owned and traded. These assets are vulnerable to theft, copyright violations, destruction, and manipulation. Traditional property crimes—such as burglary, extortion, embezzlement, and vandalism—can easily be adapted to virtual contexts. Such offenses already occur in online games, where users have stolen, copied, or destroyed virtual items, and bribed or coerced others. Similar crimes will almost certainly occur in the metaverse. While legal recognition is still developing, some early rulings have already treated these acts as legitimate crimes against property.

Acts in the metaverse can affect physical property outside it. These include acts of buying and selling physical property, preparations for the theft or destruction of physical property, or acts of defrauding others of physical property. Therefore, the metaverse can also be a site for preparations for and consummations of criminal acts against physical property.

Many countries recognize a third category of crime, alongside offenses against persons and property, known as crimes against society. These include violations that harm public order or moral standards, such as drug and weapon offenses, gambling, illegal pornography, and prostitution. In the metaverse, several of these crimes can occur in adapted forms. Gambling, pornography, and prostitution can be simulated or facilitated through avatars and virtual environments. While drugs and weapons cannot exist physically in the metaverse, their illegal sale or trade could still be coordinated within it. Thus, all these crimes may exist to some degree in virtual space.

4.2.3 Safeguarding Security in the Metaverse

A safe and secure metaverse can only be attained if both cybercrime and virtual crime risks are addressed. How can this happen? Cybercrime in the metaverse—such as hacking, data breaches, and malware attacks — often resembles traditional internet-based cybercrime and can be countered with familiar tools: secure architectures, access controls, firewalls, intrusion detection systems, security-by-design principles, vulnerability management, and user education.

Virtual crime, however, more closely mirrors real-world offenses like harassment, assault, and property damage, and requires different strategies. First, clear rules and regulations are essential to define prohibited behaviors, including not only legally criminal acts but also harmful or abusive actions that may fall outside formal law. Prevention tools can offer real-time protection. These include features like personal safety bubbles that limit avatar proximity, blocking mechanisms for text, voice, and presence, access restrictions for private spaces, and graduated response systems that progressively limit disruptive users' actions.

Monitoring tools may also be deployed to detect threats, track user behavior, and identify emotional cues that signal distress or aggression. Content moderation can help flag or remove harmful, obscene, or illegal material, thereby mitigating harassment and abuse. Strong identity and authentication systems add another layer of accountability and deterrence. Additional protective measures include employing virtual security officers and offering education and training for users and staff about rules, risks, and how to maintain a safe and respectful environment. Ultimately, all security measures must align with the fundamental rights of users, including privacy, freedom of expression, assembly, and bodily integrity in virtual spaces.

4.3 Freedom in the Metaverse

While security is essential for a thriving metaverse, the metaverse would hold little value if users were unable to exercise their basic freedoms within it. In Western ethical and legal thought, freedom is central to human flourishing and self-actualization—it allows individuals to form their own thoughts, make meaningful life choices, and pursue paths that bring them happiness. It also empowers people to contribute more fully to society by aligning their work and interests with their talents. Freedom is recognized as a fundamental human right in the UN Declaration of Human Rights, which includes specific rights such as freedom of thought, conscience, opinion, expression, peaceful assembly, and association. This section explores key challenges to freedom in virtual spaces, beginning with issues related to freedom of expression, followed by concerns around bodily integrity, freedom of movement and assembly, and finally, autonomy and protection from manipulation.

4.3.1 Freedom of Expression

The right to freedom of expression, or freedom of speech, is a core human right, allowing individuals and groups to express opinions and access information without fear of censorship or punishment. It encompasses two specific rights: the right to hold and express opinions without interference and the right to access to information. This right is worth protecting for several reasons. It helps people to gain knowledge and new perspectives and enables them to express their beliefs and opinions. Free speech also supports the functioning of democracy and serves as a check against government abuse. However, like all rights, it has limits. According to the UN's International Covenant on Civil and Political Rights (United Nations, 1966), restrictions are allowed to protect others' rights, national security, public order, or public health or morals. The Covenant also requires states to prohibit hate speech that incites discrimination or violence.

These limits apply to governments, not private platforms. Social media companies often impose stricter content moderation than would be possible under governmental oversight. Many social media companies ban speech that causes harm, spreads misinformation, or offends community standards. Commonly banned content includes threats, harassment, hate speech, pornography, incitement to violence, and misinformation.

Censorship of speech that is likely to cause harm or rights violations is not by definition unreasonable, but it requires a balancing act, in which the right to free speech is weighed against other moral considerations. Most people agree that if speech is likely to result significant harm or rights violations, it can be censored, but people tend to disagree on specific instances. While restrictions that prevent harm or rights violations may be justified, censorship of merely offensive content is harder to defend, since offense is subjective and does not always rise to the level of serious harm. Policies should weigh the benefits of expression—like personal growth, political discourse, and artistic value—against the risk of real harm.

The metaverse will include public forums and social spaces where users share text, images, video, and other content—much like today's social media platforms, and similarly subject to content moderation. While this aspect does not introduce fundamentally new challenges, the embodied and immersive nature of the metaverse does. A first challenge is that the metaverse will come with a transition from written to spoken communication, as users move from keyboard-based interaction to embodied presence, making voice the primary mode of expression. This opens the door to censorship of spoken language, either after the fact or in real time. Though real-time voice censorship tools are not yet widely available, they are emerging. For example, Intel's Bleep aims to block offensive speech in gaming by using artificial intelligence (AI) to filter harmful language during voice chat. It is easy to imagine such technology being applied to other virtual settings. However, this raises important questions: Can real-time censorship be justified, or does it represent an excessive restriction on free speech in immersive digital spaces?

A second key development will be the rise of nonverbal communication through avatars capable of expressing gestures, posture, and facial expressions. This raises questions about whether harmful or offensive nonverbal behavior, such as obscene or threatening gestures, should be censored in real time. Technical solutions could include disabling certain gestures by avatars or blurring them so others cannot see them. But are such restrictions justified? This increase in nonverbal expression will also lead to more symbolic speech—actions like gestures, clothing, signs, or coordinated behavior that convey meaning. In U.S. law, symbolic speech is protected under the First Amendment, with some exceptions. The metaverse will challenge us to decide when symbolic speech—such as extremist symbols—crosses a line, and who gets to draw that line.

A third development is the creation and integration of virtual entities—interactive simulations of people, objects, environments, or events—into virtual worlds by users or organizations. A key moral and legal question is whether these entities qualify as speech, and if so, whether they deserve protection from censorship. Legally, speech includes written, spoken, visual, and symbolic expression and often extends to artistic works and even clothing that conveys identity or belief. Should games, virtual buildings, furniture, or clothing in the metaverse count as protected speech? One useful guideline from the physical world is that if an object's primary purpose is expressive, it may qualify as speech. If its role is mainly functional or utilitarian, it generally does not. Applied to the metaverse, many virtual items—like basic tools or structures—may not be protected. However, artistic and narrative creations, including expressive games and interactive experiences, should be recognized as forms of speech and afforded appropriate protections.

4.3.2 Bodily Integrity, Freedom of Movement, and Freedom of Assembly

Let us now turn to rights are not often associated with the digital sphere. The first of these, the *right to bodily integrity*, prescribes that one's body is one's own and should not be touched, harmed, or controlled without one's consent. This right protects people from things like physical or sexual assault, being restrained, or forced medical procedures. It is a human right recognized by the UN Declaration of Human Rights and other global agreements. A key question is whether actions against avatars can violate a user's right to bodily integrity. The more users identify with their avatars—and especially when haptic tech is involved—the stronger the case for this idea. Involuntary movement constraints on avatars, forced confinement, or unwanted force feedback through haptic suits could potentially cross that line.

The *right to freedom of movement and residence* allows individuals to move freely within their country, choose where to live, leave any country, and return to their own. While this right legally applies to movement in nation-states, an advanced metaverse—with its territory, users, and rules—could be seen as having similar features. Supporting freedom of movement in the metaverse aligns with the vision of an open, inclusive space and would require removing barriers that restrict user mobility or property ownership. The *right to peaceful assembly and association*, also recognized in the Universal Declaration on Human Rights, includes the right to gather for protests, meetings, or demonstrations, and to form groups like political parties, unions, or clubs. These rights apply online as well as offline. Tech companies hosting metaverse platforms are therefore expected to respect them by allowing virtual gatherings and associations, and by limiting them only when there is a compelling, justified reason.

4.3.3 Deception and Manipulation

While we have discussed various freedoms, we have yet to address autonomy, a key concept in ethics. Autonomy is the ability to think independently, form one's own goals, and make informed choices. Without it, a person is not truly free—even if they appear to act freely—because their mind is not their own. Two major threats to autonomy are deception and manipulation, which distort rational decision-making. Deception implants false beliefs, while manipulation bypasses awareness to influence behavior without the person realizing it.

The metaverse introduces new risks in both areas. Five categories can be identified:

1. *Disinformation*: Beyond fake news on flat media, the metaverse enables immersive falsehoods—such as deepfake 3D news environments and deceptive 3D models of real-world locations or events—making disinformation more convincing than ever.
2. *Social Engineering*: Already a problem online, this will evolve in the metaverse through AI-driven bots posing as trusted individuals, or through virtual gifts that hide malware, tricking users into harmful actions.
3. *Fake Identities*: Users can impersonate others—whether specific individuals or trusted roles—to exploit trust, gain access to personal information, or manipulate others into actions they would not otherwise take.
4. *Dark Patterns:* On today's internet, dark patterns trick users into unwanted actions through misleading design. In the metaverse, the entire environment is interactive, meaning any object or space could be programmed to guide, pressure, or mislead users in subtle or manipulative ways.
5. *Personalized Reality*: Unlike the shared world we inhabit physically, the metaverse could present a tailored experience based on a user's profile—altering what they see, hear, or interact with. When users are not aware that their reality is being personalized, it becomes deceptive and undermines their autonomy by shaping their experience without their knowledge.

Together, these practices pose significant challenges to user autonomy in the metaverse, making it essential to design safeguards that support informed, independent decision-making.

4.4 Privacy in the Metaverse

Privacy is a major concern on the internet due to the large-scale collection, sharing, and use of personal data by a wide range of actors. While these concerns will carry over into the metaverse, the metaverse will also introduce new privacy issues that go beyond personal information. To understand this, it is important to revisit the broader concept of privacy.

4.4 Privacy in the Metaverse

Today, privacy is often equated with *informational privacy*—the control over how personal data is collected and used. But privacy originally referred to a wider range of protections, such as shielding one's home, family life, and private activities from intrusion. The right to privacy, first proposed by Warren and Brandeis in 1890, was about freedom from unwanted observation or interference, even when no data was recorded. With the rise of digital technology in the late twentieth century, the focus shifted almost entirely to data privacy. However, in the metaverse, non-informational privacy—such as the right to be let alone in virtual spaces—may again become central, raising concerns that are not captured by current data protection frameworks that are based on the notion of informational privacy.

Several scholars have argued that the right to privacy should protect against a wide range of intrusions into personal affairs, with personal information being just one aspect (Allen 1995; Finn et al. 2013). We follow the approach of Koops et al. (2016), who identify nine categories of personal affairs that people commonly expect to remain private (see Table 4.1). In addition, we identify three ways in which privacy can be violated (Brey 2005): physical interference, observation and surveillance, and the collection or use of personal information. For example, a private conversation can be intruded on, overheard, or recorded and shared, and these are three ways in which its privacy can be violated. Importantly on this account, personal information is both a private affair that should be respected as well as a means by which the privacy of other personal affairs can be intruded on.

4.4.1 New Privacy Risks in the Metaverse

Privacy concerns with the current internet mostly center on informational privacy—the collection and use of personal data shared through websites and communication channels. Following our account of privacy, there occasionally are other privacy

Table 4.1 Types of Privacy (after Koops et al. 2016)

Type of privacy	Type of private affair
Bodily privacy	The human body
Spatial privacy	Private spaces such as the home
Communicational privacy	Personal communication with others
Proprietary privacy	Owned property that shields access to other private affairs, like purses, briefcases, pockets and lock boxes
Intellectual privacy	The human mind, including its thoughts and opinions
Decisional privacy	Decisions of an intimate and personal nature
Associational privacy	Relations and interactions with persons, groups, and communities, especially those with a private or semi-public character
Behavioral privacy	Individual behaviors, especially those displayed in public spaces
Information privacy	Personal information

New privacy risks in the metaverse

concerns on the internet as well. For instance, eavesdropping on a video call can violate communication privacy, associational privacy, and spatial privacy, rather than informational privacy.

In the metaverse, privacy concerns will expand significantly beyond personal information. Because it is embodied and spatial, the metaverse resembles the physical world more than traditional websites. Users will interact through avatars, occupy private virtual spaces, and store personal items in digital containers. These elements introduce new vulnerabilities: users may experience violations of bodily privacy (e.g., unwanted avatar contact), spatial privacy (e.g., intrusion into private virtual spaces), or mental privacy through tools like emotion recognition or brain-computer interfaces, which may enable limited access to thoughts or feelings. Not only does the metaverse frequently involve private affairs other than personal information, the way in which these private affairs are intruded on is not just through data collection, but also through physical-like interference and through observation or surveillance. A privacy framework focused only on personal information and its collection and processing will therefore be insufficient.

This is not to say that the collection of personal data is less important in the metaverse. In fact, it is possibly more important due to the new, often *highly sensitive forms of personal data* that can be collected in it. Sensors in headsets, controllers, cameras, and wearables track eye movements, facial expressions, hand gestures, full-body motion, speech, and possibly also physiological data like heart rate, respiration rate, and electrodermal activity. Behavioral data, emotional and social data can also be inferred through biometric and user interaction data. For example, emotional states can be inferred from voice analysis, interaction habits can be inferred by monitoring user interactions, and social behavior can be analyzed through real-time tracking of user interactions with other users.

These expanded privacy risks—beyond those of both the internet and the offline world—are highly significant. The metaverse has the potential to lay bare deeply personal aspects of an individual's life, enabling the continuous collection and analysis of highly sensitive data—especially concerning the body, behavior, emotional states, and patterns of social interaction. Not only platforms and businesses, but also fellow users and hackers, will likely have powerful new tools for surveillance, observation, and manipulation, unless strong regulations are in place. Even well-meaning uses, such as personalization or safety features, could result in serious overreach if not properly regulated.

4.4.2 User Profiling, Tracking and Personalization in the Metaverse

On the internet, platform operators and businesses collect user data for purposes such as security, personalization, product improvement, and monetization. While some data is aggregated thereby becomes less privacy-sensitive, other uses involve targeted personalization, which relies on detailed personal information. In the metaverse, similar objectives will drive data collection—but the scope, sensitivity, and privacy risks will be significantly greater.

The two primary methods of data collection on the internet—user profiling and user tracking—will also be central in the metaverse. *User profiling* involves building digital representations of individuals to support identity verification, personalization, and access control. Key types include social profiling, which draws from demographics and interests, and behavioral profiling, which tracks users' activities over time to predict preferences and behaviors. These profiles enable adaptive content, personalized recommendations, and targeted advertising.

User tracking monitors real-time behavior. On the internet, it uses tools like cookies and behavioral biometrics, tracking patterns in clicks, keystrokes, or scrolling. In the metaverse, tracking will extend to full-body movement, gestures, gaze, voice, and social interaction, collected through VR headsets, motion sensors, and wearables. This allows for more detailed behavioral and psychological analysis, often without users fully realizing what is being captured.

The privacy risks of these practices are considerably higher in the metaverse due to the volume and depth of personal data involved. Platform operators may gain access to intimate details such as body language, facial expressions, emotional states, and private conversations, all of which can be analyzed to infer sensitive information—including mental health status, personality traits, or medical conditions. These inferences can then feed back into profiles used for further personalization or surveillance.

Personalization in the metaverse will take on new and immersive forms. Users will encounter personalized virtual products, services, and experiences, with content and even platform rules dynamically tailored to their skills, interests, and needs. Advertising may involve the subtle placement of products within virtual environments, appearing naturally but selected based on user profiles. Users could also be approached by virtual spokespersons—AI-driven bots—embodied virtual agents—designed to sell products or promote specific ideas. These methods risk invading privacy when they rely on detailed psychological profiling and are used to exploit users' vulnerabilities, especially if personalization is opaque or manipulative in nature. Without robust regulation and transparency, such personalization may become manipulative or coercive, challenging core principles of privacy, autonomy, and informed consent. As such, a careful balance must be struck between innovation and the protection of users' digital rights in the metaverse.

4.4.3 Surveillance Capitalism in the Metaverse

User profiling and tracking are carried out by platform operators and other online businesses to benefit their own business model, as well as for monetization involving third parties. We have found that profiling and user tracking in the metaverse could yield many new categories of data that are highly privacy-sensitive. Data analytics and biometrics can be used to build very advanced user profiles with accurate descriptions, predictions, and explanations of human behaviors, mental states, opinions, personality traits, and medical conditions.

Personal data from user profiling and tracking can be monetized in several ways. It can be sold or leased for commercial and research purposes, or analyzed to develop products like marketing tools, risk assessments, and personalized services. A major type of monetization is targeted advertising and messaging, including political messaging, which tailors content to individuals based on detailed user profiles, often having a strong psychological impact.

In the metaverse, such practices will become more immersive and potentially more invasive. New methods will include 3D interactive ads, virtual product placements, and AI-driven spokespersons. AR and VR also open the door to personalized realities—environments tailored to individual user profiles, which may deceive or manipulate users without their awareness. In this way, data monetization in the metaverse could evolve into a more intrusive form of surveillance capitalism.

Surveillance capitalism, a term introduced by Shoshana Zuboff (2019), refers to a business model in which companies collect personal data—often without consent—to sell targeted ads and services. This model operates within the broader framework of information capitalism, which Zuboff sees as the dominant form of capitalism in the digital age, driven by the creation and control of information. She argues that surveillance capitalism exploits users, erodes privacy and autonomy, and manipulates behavior for commercial gain. It also concentrates power in a few tech corporations, giving them disproportionate influence over public opinion and democratic processes, ultimately threatening the public good.

Excessive monetization and surveillance capitalism can be responded to in two ways. The first is by rejecting the business model of surveillance capitalism, arguing that its commodification of personal data is inherently harmful. This position supports strict regulation, including banning data sales, limiting targeted advertising, recognizing personal data ownership, and giving users strong rights over how their data is used.

The second view sees data monetization as potentially compatible with privacy, provided strong protections are in place. This is the stance taken, among others, by the European Union through the General Data Protection Regulation (GDPR) and Digital Services Act (DSA). These laws require clear, informed consent for data use and reuse, restrict sensitive data processing, and ban targeted advertising to minors. They also require transparency about how personalization is done and allow users to access, erase, or restrict their data.

Still, questions remain about whether such regulations are sufficient for the metaverse. Current laws often lack clear rules for new data types, analytics, and immersive environments, and they do not mandate blanket opt-outs or profit-sharing models. It is recommended that companies seeking to monetize data in the metaverse must go beyond compliance. They should ensure genuine consent, limit data use to ethical purposes, provide easy opt-outs, avoid sensitive or manipulative analytics, and consider profit-sharing with users. Responsible data practices are essential not just for legal safety, but for protecting user trust and the public good in emerging virtual worlds.

4.4.4 Privacy Protection in the Metaverse

It should be clear by now that the metaverse presents unprecedented risks to privacy. A major concern is the large-scale collection of sensitive data—such as information about users' bodies, behaviors, private environments, and spoken communications—through continuous tracking and profiling. Even more troubling is the generation of new personal data through advanced analytics, which can infer emotional states, beliefs, and personality traits to predict and potentially manipulate behavior. Given the sensitivity of such data, it is essential that its collection and use be strictly limited. Information related to health, biometrics, personality, emotions, political or ideological views, and private conversations should not be collected at all, or if collected, only for narrowly defined, justified purposes that serve a compelling user or public interest.

It is also recommended that strong privacy protections are in place for personalization and monetization. Personalization is intrusive when it draws on sensitive information—such as health, family, or sexual orientation—and presents it back to users through ads or messaging. Exploitative microtargeting in the metaverse should be avoided. Microtargeting must not use predictive analytics on users' political beliefs, personality traits, or emotional states to manipulate or bypass rational decision-making. This restriction should also apply to virtual spokespersons and product placement.

The metaverse also enables user-to-user privacy violations. Users may gain tools to spy, follow, or intrude on others' private spaces. They could eavesdrop, access private belongings, or use stealth modes and surveillance tools. To mitigate this, behavioral rules and technical safeguards must be established to protect users from one another's unwanted interference.

It is also recommended that privacy protections go beyond the processing of personal information and adopt a broader understanding of privacy that includes bodily, spatial, relational, and other dimensions. Violations can occur not only through data collection and use, but also through intrusive acts of observation and presence within (private) virtual spaces.

4.5 Equality, Fairness, and Inclusion in the Metaverse

This section explores ethical issues in the metaverse related to equality, fairness, and inclusion—interconnected values that support the vision of a society where all individuals are treated with equal respect and opportunity. *Equality* is a fundamental human right based on the belief that all people deserve the same protections and freedoms—such as privacy, security, and liberty—and should be evaluated on merit rather than personal traits like race, gender, or socioeconomic status. Equal treatment requires, among other things, protection from *discrimination*, which involves unfair treatment based on irrelevant personal characteristics. The Universal

Declaration of Human Rights identifies protected traits such as race, sex, language, religion, and national origin, and many countries have expanded this list to include sexual orientation, gender identity, disability, age, marital status, and genetic features. Individuals with these characteristics are recognized as protected groups or classes, and discrimination against them is widely prohibited in areas such as employment, housing, education, and access to services, along with related harmful behaviors like hate speech, incitement to hatred, and harassment.

Two key values closely tied to equality are *justice* and *fairness*. Justice involves the fair treatment of individuals and requires equality, as unequal treatment is inherently unjust. Both justice and fairness also rest on the idea that people should receive what they deserve—rewards for positive contributions and accountability for harmful actions—and that decisions should be made impartially, without bias or favoritism. Another related value, *diversity*, builds on these values by encouraging the inclusion of people with different identities, skills, experiences, and perspectives, which can enrich group performance and innovation. *Inclusiveness*, finally, builds on the notion of diversity and means ensuring that people of diverse backgrounds and identities are welcomed, valued, and able to fully participate in a group, organization or society. It involves creating equitable conditions, recognizing and respecting differences, and fostering a sense of belonging and mutual respect for all participants.

4.5.1 Access to the Metaverse

One of the first equality issues in the metaverse is access. As the metaverse becomes either the next version of the internet or a major part of it, it is likely to become a vital everyday service. The internet is already seen as essential because it connects people to important things like information, communication, education, healthcare, and shopping. Having access to it is key for everyone to take part in society equally. For the same reason, making sure everyone has fair and equal access to the metaverse will likely be just as important.

Scholars and policymakers increasingly recognize that digital access goes beyond having the right hardware. People also need the skills to use technology effectively, and the technology must serve their needs and interests. Communication scientist Jan van Dijk (2005; 2020) proposed a widely used model identifying four key access barriers: lack of motivation, lack of physical access, lack of skills, and lack of meaningful usage. We will use this model to explore potential access barriers in the metaverse, their impact on different social groups, and ways to overcome them.

We can largely set aside the first barrier—*motivation*—since, as seen with the internet, initial reluctance usually fades once a technology becomes essential. The second barrier, *physical access*, refers to the availability, affordability, and quality of hardware and connectivity. The metaverse requires VR/AR headsets and high-speed networks, which may be financially out of reach for low-income users. Rural users may face added challenges due to the cost of high-bandwidth infrastructure and edge computing. Physical health concerns also limit access—VR/AR is not recommended for young children and can cause side effects in some adults. Addressing these issues requires affordable hardware, expanded rural networks, and health-conscious design.

The third barrier, *skills*, relates to the ability to effectively use digital tools. On the positive side, the metaverse may lower some skills barriers that are present in the current internet by simulating real-world interaction that may make it easier to use. However, people with disabilities may face serious obstacles. For the visually impaired, the metaverse's visual nature is challenging. However, multimodal feedback, audio navigation tools (Zhao et al. 2018), and object descriptions (Gejrot et al. 2021) may offer solutions. Users with upper-body motor impairments may not be able to make the necessary head or body movements to make adequate use of head-mounted displays. However, they may benefit from brain-computer interfaces or adaptive tools like WalkinVR (2022), which adjusts virtual spaces for wheelchair users. Adaptations are also needed for people who are hearing impaired and people with speech disorders. Speech-to-text, text-to-speech and sign-language-to-text/speech translations may provide a solution. For illiterate users, accessibility can be supported by having text-to-speech options and by supporting text-free user interfaces and content. Neurodivergent users may require sensory customization in the metaverse because they often experience heightened or diminished sensitivity to sensory input, while non-dominant language speakers may require real-time translation tools.

The fourth barrier, *usage*, concerns whether different classes of people can benefit from the metaverse in areas like education, work, or civic life. Research of internet use shows that less educated users tend to use digital tools mainly for entertainment and social interaction, while more educated users use it for information, news, personal development, commercial transaction, and leisure. Without intervention, similar divides may appear in the metaverse. To promote equity, applications should be accessible to all education levels and tailored to underserved users' needs.

4.5.2 Biases in the Metaverse

We now turn to biases in the design and operation of the metaverse that may disadvantage or discriminate against users, along with ways to address them. Bias refers to systematic unfair treatment of individuals or groups. It can arise from human beliefs, institutional practices, or the design and use of technology. Here, we focus on biases in metaverse design and operations.

Design biases will be covered first. Metaverse design can be burdened by different kinds of biases in design, including physical access, skills, functional, algorithmic and representational biases. The first three are technological means by which access barriers may be posed. *Physical access bias* occurs when the technology unnecessarily limits use to certain hardware or software. *Skills bias* appears when interfaces assume users have particular skills, excluding those who do not. *Functional bias* arises when the system favors the needs of some users over others, creating barriers to use. Reducing these three types of design bias in metaverse platforms and applications is essential to ensure universal and equitable access for all users.

Algorithmic biases are biases in algorithms, especially those involved in making decisions or recommendations. Algorithms are biased if they have systematically unfair outcomes for individuals or groups, for instance based on characteristics like race or gender. These biases may occur, for example, in algorithms used in metaverse security, user profiling, personalization, and commerce. Such algorithms might—often unwittingly—treat certain users or user groups unfairly. An example is a moderation algorithm that wrongly flags speech in African American Vernacular English as offensive, leading to higher suspension rates for Black users and limiting their participation. *Representational bias*, finally, concerns how virtual worlds depict reality. Biased representations can include stereotypical portrayals of certain groups, such as racialized avatars behaving in limited or negative ways. These biases also extend to what values are emphasized—for example, a digital twin of the ocean that omits the environmental impact of local industries, favoring corporate interests over environmental concerns.

Biases also arise in the *policies and practices* of metaverse platforms and businesses. These may not be explicitly discriminatory but can still disproportionately harm vulnerable groups. For instance, weak protections against harassment may particularly affect marginalized users. Similarly, supporting Christmas celebrations while ignoring those of other religions unfairly favors one group.

To reduce bias, designers can adopt frameworks like *Design for Values* (Van Der Hoven and Manders-Huits 2020), which promotes fairness, universal access, and ethical impact. This includes addressing algorithmic bias, ensuring usability for all, and building fairness into every design stage. Other approaches—such as *universal design* (Steinfeld and Maisel 2012), inclusive design (Keates and Clarkson 2004; Microsoft Design 2016), and algorithmic fairness (Pessach and Shmueli 2020)—can further support these goals. Involving diverse users in design processes ensures their needs are considered. Platform policies should prioritize equity, and companies should build diverse, inclusive teams to help prevent unconscious bias from shaping design and decision-making.

4.5.3 User to User Discrimination

Discrimination in the metaverse can occur through user-to-user interactions, including harassment, hate speech, racist or sexist content, and exclusionary behavior—problems already widespread online. The metaverse may worsen these issues by enabling more embodied and immersive forms of abuse. A key challenge is the spread of *hateful content*, which in the metaverse will include not just text but also spoken language, nonverbal gestures, symbolic acts, and 3D imagery. Strong moderation systems and penalties will be essential to protect users from such behavior.

Another concern is *discriminatory harassment*—the harassment of individuals based on race, gender, religion, sexual orientation, or disability. Unlike text-based harassment online, harassment in the metaverse can take physical and interactive forms, making it feel more immediate and threatening. For example, a user might program their avatar to repeatedly invade another user's personal space or simulate unwanted touching or follow and verbally harass someone based on their identity in a shared virtual environment. A third risk involves *hate crimes* other than hate speech and discriminatory harassment. Hate crimes are crimes motivated by prejudice in which a victim is targeted because of their membership of a protected social group. Hate crimes in the metaverse can include physical and sexual assault, mate crime, and damage to property. These may become common if users feel anonymous and unaccountable.

Prevention and prosecution of user-to-user discrimination in the metaverse will require clear platform policies, proactive moderation, user identification policies, and dedicated enforcement. VR and AR offer a promising way to reduce discrimination by fostering empathy through immersive experiences that let users "step into the shoes" of others—across race, gender, or ability. These tools are already used in education and corporate training, with studies confirming their impact on reducing bias (e.g., Banakou et al. 2016; Herrera et al. 2018; Salmanowitz 2018). Metaverse platforms could provide such experiences for user education, while games and other applications could promote understanding by encouraging exploration of diverse perspectives.

4.6 Property Rights in the Metaverse

In the metaverse, various types of goods will play a central role, with goods being created, bought, sold, rented, lent, and stored. While *physical goods*, such as cars, houses and furniture, cannot be directly present in the metaverse, they can still be traded within it, and their ownership can be recorded—for example through blockchain technologies. *Digital goods*, by contrast, not only can be traded and registered, but also exist and function within the metaverse. They can be created, owned, used, displayed, and exchanged in immersive environments. Digital goods are items that exist in digital form, such as software, eBooks, digital music and videos, images, educational materials, games, and cryptocurrencies. Most have economic value and are owned by their creators or buyers. As IP, they are typically protected by copyright and other IP laws.

While digital goods have existed for decades, *virtual goods* are a newer category. These are digital assets that exist specifically within virtual environments like the metaverse or other online worlds. Examples include virtual clothing, furniture, currencies, and real estate. Additionally, *virtual services*—such as virtual tourism, education, training, and even virtual sex work—can hold significant economic value.

Some virtual worlds host elaborate virtual economies, where virtual goods have sold for thousands of U.S. dollars, reflecting their real-world financial and social significance. The metaverse will likely support several economies, including virtual economies, digital economies, and economies of traditional, physical goods and services, which are offered through e-commerce.

The right to own property is a fundamental human right, as it shapes people's access to valuable and beneficial resources—such as land, housing, goods, and materials. Property ownership plays a crucial role in securing individual rights, freedom, and well-being. For this reason, how property rights are defined and managed in the metaverse will be essential to protecting other rights and supporting users' overall quality of life.

We will discuss three key social and ethical challenges for property rights governance in the metaverse. These are the protection of property rights for digital and virtual property, the relation between property rights and the common good, and ethical challenges of cryptocurrencies and NFTs.

4.6.1 Protecting Property Rights in the Metaverse

Ownership is commonly defined as a "bundle of rights," particularly in U.S. law. One influential account by Honoré (1961) outlines key rights that ownership entails, such as the rights to possess, use, manage, earn income from, destroy, and transfer the good, along with the right to remain the owner over time. This framework also applies to intellectual property—creations of the mind like designs, ideas, and artworks. As Dusollier (2020) notes, intellectual property holders have the exclusive rights to control, use, manage, and monetize their creations, as well as to prevent others from reproducing or distributing them without permission.

Property rights can be violated in various ways, often through criminal acts that deprive owners of possession, control, income, or use of their goods. Common violations include theft, fraud, vandalism, intellectual property infringement, trespassing, and encroachment. These actions also occur with digital and virtual goods. They are acts that are usually clear-cut, in that they are recognizable as property rights violations, and can be categorized as both unethical and illegal.

It is the responsibility of legislators, law enforcement, platform operators, and online businesses to safeguard property rights for digital and virtual goods—both online and in the metaverse. However, current protections are often insufficient. One major issue is that many countries lack specific legal frameworks for virtual goods, unlike other digital assets. Because interoperability between virtual worlds remains limited, these goods are typically restricted to individual platforms and governed by EULAs instead of property law. As a result, users are granted only limited, revocable rights rather than full legal ownership, with little recourse if access is lost or items are removed. A rare exception occurred in a lawsuit against Linden Labs, where a user successfully contested the loss of virtual land in *Second Life* (DaCunha 2010). The case prompted the platform to revise its terms and clarify

4.6 Property Rights in the Metaverse

that virtual ownership was not legally binding but instead a temporary and revocable right.

This situation raises ethical concerns, especially given the high value of some virtual goods. Legal scholars debate whether user rights should be protected under contract, consumer protection, or property law—with property law offering broader protections like the right to transfer, sell, or seek compensation for loss. While NFTs are seen by some as a solution to ownership problems, they lack consistent legal recognition and remain tied to platform terms and existing legal frameworks. Thus, current protections for virtual property are inadequate, and there is a moral obligation for sellers and platforms to clarify ownership limits and provide better safeguards.

A second issue is the adequacy of legal enforcement mechanisms for IP rights in relation to digital and virtual goods. Digital goods like music, software, and videos are often protected by IP rights, giving creators exclusive control over how their work is used, sold, and shared. Virtual goods can also fall under IP law, especially when they involve original user-generated designs. However, enforcement is challenging, as digital goods are easily copied, leading to piracy, plagiarism, and unauthorized use. Tools like DRM, watermarking, and NFTs offer some protection for digital and virtual assets, but legal safeguards remain limited. NFTs are often tied to digital originals but usually do not confer copyright, creating legal ambiguity—especially given their volatility. Users who create virtual goods may also lose rights under platform EULAs, which often assert ownership or impose licensing restrictions. Although U.S. and EU laws can support user rights, enforcement is frequently undermined by such contractual terms, particularly when users depend on platform-provided tools and environments.

AI-powered 3D creation tools add further complexity by generating virtual objects from text or images, raising new questions about authorship and copyright. In the U.S., AI-generated content generally lacks copyright unless the human creator directly controls the result. The EU takes a more flexible approach, using a four-step test to assess if human input is sufficient for authorship. As AI-generated content becomes more common in the metaverse, legal recognition of user rights will be critical. If such content lacks copyright protection, it may default to the public domain, freely usable unless restricted by platform terms. How copyright law responds to AI-assisted creation will ultimately shape control over the metaverse—whether it lies with users or platforms.

A third issue with property rights in the metaverse is the lack of adequate protection against fraud, deception, and market manipulation. Buyers of digital and virtual goods risk overpaying for inflated assets, receiving misleading products, or facing unfair access. Poor regulation and unclear pricing, especially for NFTs, make exploitation easier, while tactics like "pump and dump" schemes increase financial risk. Stronger regulation and ethical platform practices are urgently needed. Deceptive advertising is also a concern, particularly for physical goods marketed through virtual simulations. These previews may misrepresent real products, requiring clear standards to ensure accuracy and protect consumers.

In conclusion, there are serious inadequacies in legal systems and platform protections for digital and virtual property rights, and if they persist, then property rights will not be adequately protected in the metaverse. Specific issues are the lack of adequate property rights frameworks and protection mechanisms for virtual goods and for IP rights in relation to digital and virtual goods, and the lack of adequate protections against fraud, deception and market manipulation in relation to such goods.

4.6.2 Ownership Models and the Common Good

Property ownership in the metaverse can also be assessed with respect to the common good, going beyond individual rights to explore how ownership structures can promote societal well-being, fairness, and democratic values. Most philosophical and legal theories of property hold that property rights should be arranged not only to respect individual interests but also to support broader social benefits. In theories of property, various ownership models are distinguished, such as private, public and collective ownership, and centralized and decentralized ownership, and for-profit and non-profit ownership. These ownership models have different implications for the common good.

What kind of ownership model in the metaverse best serves the common good? To answer this, it is important to distinguish between *platform ownership*—control over the underlying digital infrastructure—and *virtual world ownership*—control over virtual land, goods, and assets. Ownership structures for both platforms and virtual worlds can be *centralized*, where a single entity holds control over the platform or virtual world, or *decentralized*, where ownership is distributed among multiple private individuals, organizations, or collectively managed through models like cooperatives or decentralized autonomous organizations (DAOs). In addition, ownership can of each can be *private* (by companies and individuals), *public* (by government entities and public institutions) and *collective* (by user communities). The implications of each model for fairness, participation, innovation, and user rights are critical when evaluating their alignment with the common good.

In current virtual worlds, ownership models vary widely, from private commercial platforms like Second Life to community-governed ones like Decentraland. Many virtual worlds are centrally owned and operated, with the platform operator as the sole owner of both the platform and virtual goods in the virtual world. This is a *closed* ownership model. Some virtual worlds, like Second Life and Roblox, are centrally owned, but the virtual world layer is *semi-open*: users, collectives, and public entities can buy virtual land, real estate, and other items in it, though ownership remains limited and not absolute. Other virtual worlds, such as Somnium Space and Cryptovoxels, still have centralized ownership of the platform, but they approach an *open* metaverse model for the virtual world layer, in which users can create, buy and use virtual assets and have real, enforceable rights over them. Virtual world platforms can also be decentrally owned and operated, by groups of private owners who maintain private claims for part of the platform, or by collectives who engage

4.6 Property Rights in the Metaverse

in collective ownership. Examples are Decentraland and The Sandbox, which combine collective platform governance via DAOs with private ownership of virtual land and assets by users.

The interests of virtual property owners, creators and users are obviously best served with open ownership models for the virtual world layer, as these give them an opportunity to own virtual property and provide the best protection for their property rights. As for the property model for the metaverse platform: centralized ownership by a for-profit owner has the advantage of efficient decision-making and fast, market-driven innovation. However, their commitment to the social good may be wanting. Centralized metaverse platforms owned by governments, public institutions, or non-profits may better align with user interests and promote the common good due to public accountability. However, they can be less efficient, innovative, or responsive than private firms, and may be vulnerable to political influence. Public-private partnerships could balance innovation with public accountability, though they may introduce complexity and reduce transparency.

Decentralized and collectively owned models promote user freedom and democratic participation but can face barriers, such as complex decision-making, exclusion of less-skilled users, and limited adaptability. Still, these models provide protection from centralized authority overreach. Ideally, the metaverse should support a mix of private, public, and collective property, ensuring access, innovation, and fairness. Open virtual worlds can be supported by both centralized and decentralized platforms, as long as users retain meaningful rights. While collective ownership of platforms may enhance democratic control, well-designed regulations requiring private platforms to support openness and user rights may offer a practical balance, combining the strengths of different models while minimizing their weaknesses.

4.6.3 Cryptocurrencies and NFTs

Cryptocurrencies and NFTs are two new digital asset classes, enabled by blockchain, that are often associated with the metaverse and that could have an impact on its development. In this section, we will consider both and subject them to an ethical assessment.

Let us first consider ethical issues concerning cryptocurrencies. Cryptocurrencies are decentralized digital currencies secured by cryptography and managed on blockchain networks. They allow users to make peer-to-peer transactions without intermediaries and offer benefits such as anonymity, global accessibility, faster payments, and lower transaction costs. These traits make them well-suited to virtual environments, where frequent microtransactions, anonymity, and decentralized services are common. Cryptocurrencies also support other blockchain-based systems like smart contracts, decentralized finance (DeFi), and digital identity management.

However, their advantages are offset by significant risks (Dierksmeier and Seele, 2016). Despite their security features, cryptocurrencies are still vulnerable to hacking, scams, and fraud, and lack clear accountability structures due to minimal regulation. Their anonymity and decentralized nature also make them attractive for illegal activities such as money laundering, tax evasion, and ransomware payments. Ethical concerns include increasing inequality, as early adopters have gained immense wealth while others face losses, and barriers to entry remain high for those without resources or digital literacy. Additionally, many cryptocurrencies have high environmental costs due to the energy-intensive mining process.

These problems are being recognized within the crypto industry, and some are being addressed through technological improvements and evolving regulations. Still, greater oversight could reduce decentralization and user control—key advantages of cryptocurrencies. If crypto is to become the dominant currency in the metaverse, it must be made more secure, accountable, and inclusive. Criminal misuse must be prevented, inequality addressed, and environmental impact significantly reduced for it to ethically support metaverse economies.

Let us now turn to NFTs. Blockchain technology enables the creation of tokens, digital units of value recorded on a blockchain. These include *fungible tokens*, like cryptocurrencies (e.g., Bitcoin), which are interchangeable, and *non-fungible tokens* (NFTs), which are unique and represent specific digital assets such as artwork, music, or virtual real estate. NFTs are verified through a digital signature and can prove ownership and authenticity of virtual goods. However, ownership of an NFT usually does not include copyright, unless explicitly transferred. This means users may own a digital asset without having rights to reproduce or distribute it.

NFTs are used to represent and trade virtual assets, support user identity, or unlock access to exclusive experiences. They enable new revenue streams, promote interoperability between platforms, and foster community engagement. For example, NFTs from the Bored Ape Yacht Club provide club membership benefits beyond the digital artwork itself. However, despite their promise, NFTs raise serious ethical concerns. They have been criticized for volatility, speculative hype, lack of clarity in intellectual property rights, and enabling fraud or manipulation. Philosopher Catherine Flick has argued that NFT use cases have so far failed to meet ethical standards, often prioritizing profit over social or environmental benefit, and contributing to inequality, privacy risks, and exclusion of those without access to capital or technical expertise (Flick 2022).

To fulfill their potential, NFTs must move beyond speculative hype. The industry needs more responsible actors, better product quality, and clearer accountability. Regulation may offer consumer protections, but the decentralized nature of blockchain complicates enforcement. Thus, self-regulation by developers and marketplaces is essential for establishing ethical and sustainable practices. If guided properly, NFTs could still become a valuable tool for ownership, creativity, and community in the emerging metaverse.

4.7 Conclusion

The metaverse is envisioned as a space where users work, socialize, and engage in commerce, creativity, and leisure—mirroring many aspects of the physical world. Safeguarding rights in this environment is essential to ensure that users maintain control over their data, identity, and digital property, and are protected from exploitation, discrimination, and abuse. Without proper development, management, and regulation, the metaverse could pose serious threats to fundamental rights such as security, privacy, freedom, equality, and property. To prevent this, clear strategies must be implemented to protect these rights and effectively mitigate potential infringements.

The right to security may be compromised by both traditional and novel cybersecurity threats and emerging virtual security risks targeting people and assets in virtual worlds. While standard tools can be used to address cybersecurity, threats to virtual security—like harassment or assault—require behavior-focused measures such as clear rules, real-time protections, monitoring, moderation, identity systems, and user education. Freedom rights that could be at risk in the metaverse include freedom of and expression, bodily integrity, freedom of movement, freedom of assembly, as well as the right to autonomy. We examined the unique ways these rights may be threatened in the metaverse and outlined mitigation strategies, including regulatory measures, self-regulation by platforms, and design approaches.

Privacy risks in the metaverse can be profound, we found. The metaverse poses a serious risk of collecting and generating highly sensitive personal data through tracking, profiling and analytics, including inferences about emotions, beliefs, and personality. Advanced personalization using biometric and behavioral data can fuel targeted immersive ads and political messaging that manipulate users without full awareness. There are also threats stemming from the monetization of personal data as well as user to user privacy violations. We reviewed the privacy protections that are needed to mitigate these threats to privacy.

The metaverse should also support equality, fairness, and inclusion. Equal access to the metaverse requires that physical access, skills, and usage barriers that plague different user groups are overcome. Biases in metaverse design, like algorithmic and representational biases, can also lead to inequality and unfair outcomes for social groups, and we discussed design methods for mitigating them, like Design for Values. User to user discrimination in the metaverse is another concern that we discussed mitigation strategies for. Property rights in the metaverse need to be rethought and strengthened. Current protections for digital and virtual assets are inadequate, highlighting the need for legal reform and more robust platform governance. In considering how ownership structures can serve the public interest, it was concluded that an open metaverse—where users have meaningful control over their assets—best supports shared social and ethical goals. We also examined the ethical challenges associated with cryptocurrencies and NFTs as metaverse assets, finding that stronger regulation, self-regulation, transparency, and accountability are badly needed to ensure beneficial outcomes.

References

Allen, Anita L. 1995. Privacy in Health Care. In *Encyclopedia of Bioethics*, ed. Warren T. Reich, 2064–2073. New York, NY: Simon and Schuster Macmillan.

Banakou, D., P. D. Hanumanthu, and M. Slater. 2016. Virtual Embodiment of White People in a Black Virtual Body Leads to a Sustained Reduction in Their Implicit Racial Bias. *Frontiers in Human Neuroscience* 10:601. https://doi.org/10.3389/fnhum.2016.00601.

Brey, Philip. 2005. The Importance of Privacy in the Workplace. In *The Ethics of Workplace Privacy*, ed. Sven Ove Hansson and Elin Palm, 97–118. Brussels: Peter Lang Verlagsgruppe.

DaCunha, Nelson. 2010. Virtual Property, Real Concerns. *Akron Intellectual Property Journal* 4 (1): 35–66.

Dierksmeier, C., and P. Seele. 2016. Cryptocurrencies and Business Ethics. *Journal of Business Ethics* 152 (1): 1–14. https://doi.org/10.1007/s10551-016-3298-0.

Dusollier, Séverine. 2020. Intellectual Property and the Bundle-of-Rights Metaphor. In *Kritika: Essays on Intellectual Property*, ed. Peter Drahos, Gustavo Ghidini, and Hanns Ullrich, 146–179. Cheltenham: Edward Elgar Publishing.

Finn, Rachel L., David Wright, and Michael Friedewald. 2013. Seven Types of Privacy. In *European Data Protection: Coming of Age*, ed. Serge Gutwirth, Ronald Leenes, Paul de Hert, and Yves Poullet, 3–32. Dordrecht: Springer.

Flick, C. 2022. A Critical Professional Ethical Analysis of Non-Fungible Tokens (NFTs). *Journal of Responsible Technology* 12:100054. https://doi.org/10.1016/j.jrt.2022.100054.

Gejrot, Emil, Sara Kjellstrand, and Susanna Laurin. 2021. *New realities: Unlocking the potential of XR for persons with disabilities.*. Commissioned by Facebook under Agreement Research on Accessibility Benefits of AR and VR Technology. https://www.funka.com/contentassets/0efed77304a3467aa7fdd7e181c7b47b/new_realities_xr_for_pwd.pdf. Accessed 29 Mar 2025.

Herrera, F., J. Bailenson, E. Weisz, E. Ogle, and J. Zaki. 2018. Building Long-Term Empathy: A Large-Scale Comparison of Traditional and Virtual Reality Perspective-Taking. *PloS One* 13 (10): e0204494. https://doi.org/10.1371/journal.pone.0204494.

Honoré, Tony. 1961. Ownership. In *Oxford Essays in Jurisprudence: A Collaborative Work*, ed. Anthony G. Guest, 107–147. New York: Oxford University Press.

Keates, Simeon, and John Clarkson. 2004. *Countering Design Exclusion: An Introduction to Inclusive Design*. London: Springer.

Koops, Bert-Jaap, Bryce C. Newell, Tjerk Timan, Ivan Skorvánek, Tomislav Chokrevski, and Maša Galič. 2016. A Typology of Privacy. *University of Pennsylvania Journal of International Law* 38:483–575.

Microsoft Design. 2016. *Inclusive Design*. https://www.microsoft.com/design/inclusive/. Accessed 29 Mar 2025.

Pessach, Dana, and Erez Shmueli. 2020. Algorithmic Fairness. arXiv. https://arxiv.org/abs/2001.09784. Accessed 29 Mar 2025.

Salmanowitz, N. 2018. The Impact of Virtual Reality on Implicit Racial Bias and Mock Legal Decisions. *Journal of Law and the Biosciences* 5 (1): 174–203. https://doi.org/10.1093/jlb/lsy005.

Steinfeld, Edward, and Jordana Maisel. 2012. *Universal Design: Creating Inclusive Environments*. Hoboken, NJ: John Wiley & Sons.

Tapscott, A. 2024. Meta's Splashy Orion Glasses Are Just Another Tool for Big-Tech Dominance Over Our Future. New York Post, October 5. https://nypost.com/2024/10/05/opinion/metas-orion-glasses-are-another-tool-for-big-tech-dominance/. Accessed 1 Apr 2025.

United Nations. 1966. International Covenant on Civil and Political Rights. OHCHR, December 16. https://www.ohchr.org/en/instruments-mechanisms/instruments/international-covenant-civil-and-political-rights. Accessed 29 Mar 2025.

Van der Hoven, Jeroen, and Noemi Manders-Huits. 2020. Value-Sensitive Design. In *The Ethics of Information Technologies*, ed. Keith Miller and Mariarosaria Taddeo, 329–332. London: Routledge.

Van Dijk, Jan. 2005. *The Deepening Divide: Inequality in the Information Society*. London; Thousand Oaks; New Delhy: Sage.

Van Dijk, Jan. 2020. *The Digital Divide*. Cambridge, UK; Medford, USA: Polity Press.

Zhao, Y., C.L. Bennett, H. Benko, E. Cutrell, C. Holz, M.R. Morris, and M. Sinclair. 2018. Enabling People with Visual Impairments to Navigate Virtual Reality with a Haptic and Auditory Cane Simulation. In *Proceedings of the 2018 CHI Conference on Human Factors in Computing Systems*, 1–14. https://doi.org/10.1145/3173574.3173690.

Zuboff, Shoshana. 2019. *The Age of Surveillance Capitalism: The Fight for a Human Future at the New Frontier of Power*. New York: PublicAffairs.

Chapter 5
Will the Metaverse Benefit Humans and Society?

> "It's just an online social entertainment experience in a real-time 3D setting. You and your friends, going around having fun together, in a 3D world." [Webster (2023)]—Tom Sweeney, CEO Epic

Abstract This chapter of *The Metaverse: A Critical Assessment* explores the potential benefits and risks of the metaverse for individuals and society at large. At the individual level, the central question is whether the metaverse can enhance human well-being; at the societal level, whether it can contribute to the public good by strengthening key social institutions such as education, healthcare, and democratic governance, as well as supporting civic life and environmental sustainability. The first section focuses on individual well-being, examining how metaverse use may both support and undermine aspects such as pleasure, health, achievement, meaning, personal growth, and social connection—with particular attention to risks for mental health and social well-being. The next section assesses the metaverse's potential to serve the public good by evaluating its impact on domains like work, education, healthcare, public discourse, and civic engagement. It also considers whether the social and political structures within the metaverse can promote social cohesion and resilience and democratic governance. The final section addresses the metaverse's environmental implications, recognizing sustainability as a critical component of the public good.

Keywords Metaverse · Well-being · Harm · Mental health · Addiction · Social Well-being · Friendship · Democracy

5.1 Introduction

In this chapter, we will examine benefits and risks of the metaverse for individual human beings and for society at large. For humans, the key question is whether the metaverse will enhance their well-being. Without improvements in individual well-being, no technology can truly be said to benefit humans at a basic level. In the next section, it will be examined how the metaverse could benefit individual well-being and how it could harm it.

For society, the key question is whether the metaverse will contribute to the social good. While the well-being of citizens is a core component of the social good, the social good as a concept is broader. It refers to the set of conditions, systems, and principles that benefit society as a whole, not just individuals in isolation. It refers to the quality of institutions like education and healthcare, and the quality of economic opportunity, civic life, and democratic participation. In Sect. 5.3, we will consider both the quality of social systems in the metaverse, and how the metaverse might impact society more broadly. Section 5.4, finally, will focus on the potential impact of the metaverse on environmental sustainability, which is also a vital part of the social good.

5.2 Well-being in the Metaverse

At the end of the day, people will use the metaverse only if it truly improves their lives. Will it enhance their well-being by offering new opportunities, enriching experiences, and meaningful social connections? Or will it diminish well-being by fostering addiction, social isolation, exploitation, and the erosion of privacy, autonomy, and real-world relationships? In this section, we explore both the potential positive and negative impacts of the metaverse on well-being and examine the responsibility of platform operators to safeguard and promote the well-being of their users.

Well-being, alternatively called welfare or quality of life, concerns the goodness of one's life. It is sometimes equated with happiness, but it has also been claimed that being happy is only part of what makes a life good. Well-being is not a right, as it cannot be guaranteed by others and remains primarily an individual responsibility. People have a right to be protected from serious harms that undermine their ability to secure their own well-being, and they should be able to have the resources necessary for having a good life.

On an optimistic view, the metaverse will be a new guarantor of well-being and happiness. It is a utopia of unlimited choice and self-realization, where users can be whoever they want and fulfill every desire without harming others. However, this vision is deeply flawed. The metaverse is not an isolated playground but a shared environment that demands compromise, mutual respect, and navigation of both social and technical limitations. From a pessimistic standpoint, the metaverse cannot deliver genuine happiness or fulfillment because it is a simulated reality, inherently lacking authenticity. While this view rightly warns against overvaluing virtual objects or mistaking fiction for reality, it overlooks the fact that many metaverse experiences—especially those involving real human interaction—can be meaningful, authentic, and enriching. Both these views are overly simplistic; the metaverse holds real potential to enhance well-being, but only within limits that demand critical reflection, purposeful engagement, and a healthy balance between virtual experiences and real-world life.

5.2.1 Key Aspects of Well-Being and the Metaverse

Various theories of well-being have been proposed, in the fields of philosophy, psychology, and economics (Brey 2012). There is ongoing debate about what well-being truly is and whether it can be defined in a way that applies universally. Without going into this debate, we will discuss certain elements of well-being that many theories of well-being can agree on. These include pleasure, physical health, mental health, achievement, meaningfulness, personal growth, and friendship and social well-being. In this section, we will examine for each how the metaverse is likely to impact their realization.

One key aspect of well-being is *pleasure*, which includes both short-term enjoyment and long-term life satisfaction. The metaverse has the potential to provide rich and pleasurable experiences, ranging from immersive adventures to engaging social interactions. However, its role will often extend beyond leisure, blending into the routines of real life—facilitating office work, education, commerce, and other everyday activities. While these functions can enhance convenience and enjoyment, deeper and more lasting forms of well-being depend on factors such as emotional security, meaningful relationships, personal growth, and a sense of purpose. Whether the metaverse can truly support these dimensions remains uncertain, and its impact will largely depend on how it is designed, used, and integrated into human life.

Physical health is another major component of well-being. Prolonged, sedentary use of the metaverse may pose health risks, including eye strain, poor posture, and muscle atrophy. However, the growing integration of active virtual reality (VR) technologies and immersive fitness platforms presents opportunities to encourage physical activity—particularly for individuals with limited mobility or access to traditional exercise environments. In addition, the emerging "medical metaverse" is expanding the scope of healthcare, enabling virtual consultations, remote monitoring, rehabilitation programs, and interactive health education. These innovations highlight the metaverse's potential not only to entertain but also to support physical well-being in meaningful ways.

Achievement supports well-being because it contributes significantly to self-esteem and satisfaction. The metaverse can support achievements in areas such as education, career, finance, and hobbies. However, some meaningful achievements are tied to the physical world and are not replicable in the metaverse—like raising children, excelling in traditional sports, or mastering hands-on real-world crafts. As with current digital technologies, there is also a risk of over-investing in superficial digital achievements, such as social media metrics or gaming scores, which may offer limited real-world value.

Meaningfulness is widely recognized by well-being scholars as an important aspect of a good life, with psychologist Martin Seligman (2011) naming it as a core element. It gives people a sense of purpose and connects them to something greater than themselves. Meaningfulness can arise from community service, activism, spirituality, personal relationships, and fulfilling work or hobbies. Many of these pursuits can, in principle, be supported in the metaverse. However, the metaverse might also encourage shallow activities that lack real purpose, and its use could negatively impact mental health, reducing one's capacity to pursue meaningful goals. These concerns will be explored further below.

Many people strive to become better by developing their talents and realizing their potential—intellectually, socially, emotionally, and morally. This process, known as *personal growth*, supports well-being by enhancing capabilities and decision-making. The metaverse has the potential to support personal growth through immersive experiences, education, and exploration of new identities and practices. Realizing this potential will require intentional design. If the metaverse prioritizes only entertainment and commerce, growth opportunities may be limited. However, if it also supports experiential learning and personal development, it can play a valuable role in fostering growth. This requires realistic simulations, a wide range of experiences, and thoughtful design that helps users avoid choices that undermine their development.

5.2.2 The Metaverse and Mental Health

Mental health is a key aspect of well-being. Its absence almost guarantees a reduced quality of life. Mental health is not merely the absence of mental disorders. Mental health refers to a person's capacity to cope with life's challenges, manage stress, and maintain emotional well-being. It includes the ability to develop one's potential, work productively, and contribute meaningfully to one's community and society.

The impact of internet use on mental health offers valuable insights into how the metaverse might affect mental well-being. Studies on the internet's role in well-being and mental health have examined its correlation with general happiness, emotional well-being, self-esteem, and social connectedness, as well as with negative outcomes like anxiety, depression, loneliness, and addiction. While media often report strong negative effects—particularly from social media—empirical evidence is more nuanced. Orben (2020), reviewing studies on teens, found only a small average negative link between social media use and well-being, and weak evidence for correlations with depression and anxiety. In another review, Valkenburg (2021) also found mixed results, with some studies showing links between social media use and depressive symptoms, while others showing none.

Some studies also report positive correlations between internet use and mental health. Orben (2020) cites research showing social media use can boost well-being, social connection, and reduce loneliness among teens. Forsman and Nordmyr (2017) found similar results among older adults, linking internet use with improved mental health, social support, and lower isolation. Valkenburg (2021) argues that differences among users explain these mixed outcomes. One study found 20% of users experienced positive effects, 20% negative, and 60% no effect. Importantly, frequent or intense internet use is consistently linked to lower overall well-being and increased symptoms of anxiety and depression. These findings suggest that while internet use may support mental health for many, it can have harmful consequences for a significant minority—particularly those with preexisting mental health issues, negative online experiences, or patterns of excessive use.

Like the internet, the metaverse may improve mental health for many by expanding resources, offering new forms of connection, and building social capital. These benefits—such as enhanced mood, self-esteem, life satisfaction, and a sense of purpose—could be even greater than those seen with today's internet, given the metaverse's more immersive nature. However, a significant number of users may experience negative mental health impacts, including depression, anxiety, low self-esteem, and social isolation. These may stem from preexisting mental health issues, harmful social interactions, lack of connection, or compulsive use—issues already known from internet studies but possibly intensified by the immersive quality of the metaverse.

A key concern is the embodied nature of the metaverse, which may increase risks to self-worth and self-esteem by amplifying issues already present on the internet, especially on social media. Platforms like Instagram and Snapchat promote self-presentation and self-promotion through curated images and videos, encouraging users to portray idealized versions of themselves and their lives. These practices, supported by filters and editing tools, can lead to harmful social comparison and pressure to seek validation through likes and positive feedback. While most users are unaffected, studies show a significant minority experience reduced self-esteem (Cingel et al. 2022), particularly women, adolescents, and young adults (Araigy 2018; Lockhart 2019).

This curated "Instagram reality" has been linked to body dissatisfaction and body dysmorphic disorder. Meta-analyses (Vandenbosch et al. 2022; Saiphoo and Vahedi 2019) show small but significant correlations between social media use, selfie behavior, and negative body image—especially among younger users. These risks may be heightened in the metaverse, where users have full-body, real-time presence through photorealistic avatars. Enhanced possibilities for interaction and self-modification—such as idealized appearances or AI-augmented personalities—could intensify self-presentation pressures. As a result, the negative impact on self-esteem and body image may exceed that of current social media platforms.

Another significant mental health concern is the risk of *metaverse addiction*, which has become a prominent topic in discussions about the technology (Floridi 2022; Reed and Joseff 2022). The concern is that some users may become dependent on the metaverse for its pleasure and instant gratification, leading to excessive use or escapism from real-life responsibilities. If these patterns emerge, they could significantly harm users' mental health and well-being, as well as negatively impact their relationships and daily functioning.

Addiction is defined as a compulsive, chronic need to engage in a behavior or use a substance despite harmful consequences. It can lead to physical, psychological, and social problems, and withdrawal often causes anxiety, irritability, or other symptoms. Addiction is recognized as a neuropsychological disorder involving brain changes that reinforce the behavior. One influential explanation is the dopamine theory, which suggests addiction stems from overstimulation of the brain's reward system (Nutt et al. 2015), though other theories exist. Addictions are typically classified as substance or behavioral (Robbins and Clark 2015). While substance addictions are long recognized, behavioral addictions—such as internet, gaming, or social media addiction—are gaining attention, though most are not yet formally recognized in major diagnostic systems.

Alleged internet-related behavioral addictions include internet gaming disorder, social media addiction, pornography addiction, and broader concepts like digital and internet addiction. Among these, *internet gaming disorder* has the most formal recognition, appearing in the DSM-5 classificatory system and recognized by the WHO. Social media and online pornography addictions have substantial scientific backing, with research linking them to reduced self-esteem, body dissatisfaction, and mental health issues. Broader terms like digital and internet addiction face criticism for being too vague, as addiction often relates to specific content.

Regardless of whether certain internet-related behaviors are officially classified as addictions, their negative effects are well-documented. Studies show a significant portion of the population engages in problematic social media use (Bányai et al. 2017; Paakkari et al. 2021), linked to poor work performance, unhealthy relationships, low life satisfaction, anxiety, depression, and negative self-image (Sun and Zhang 2021; Cingel et al. 2022; Ryding and Kuss 2020). Similar concerns apply to online pornography, with research associating its problematic use with sexual dysfunction, relationship issues, mood disorders, and substance abuse (de Alarcón et al. 2019; Sniewski et al. 2018). Data from the U.S. and Sweden suggest high rates of problematic and addictive pornography use. Internet gaming disorder affects millions globally, with compulsive gaming interfering with daily life, self-care, and emotional well-being (Camilleri et al. 2021; Stevens et al. 2021). In addition, many platforms have been accused of deliberately incorporating addictive features—such as infinite scroll, variable rewards, and autoplay—to encourage compulsive use. This approach, widely seen in both social media and gaming, is known as *addiction by design* (Schüll 2014).

The metaverse is expected to include many potentially addictive activities already found on the current internet—such as gaming, social media, gambling, pornography, shopping, and work—but in a far more immersive format, which could increase their addictive potential. Unlike traditional screen-based experiences, the metaverse surrounds users in fully embodied environments, engaging more senses and intensifying stimulation. Some have warned that the metaverse could be the "fentanyl of the internet," making online behaviors far more addictive (Sternlicht and Sternlicht 2022).

While research is still limited, some studies suggest immersive technologies may indeed increase addiction risk. Barreda-Ángeles and Hartmann (2022) found that while overall VR addiction was not higher than gaming addiction, feelings of embodiment predicted more compulsive use. Similar findings appear in studies of gaming and VR, where presence correlates with stronger emotional arousal and enjoyment—both linked to addictive behavior (Stavropoulos et al. 2019; Elsey et al. 2019). Moreover, when high-reward structures like those in gambling or gaming are combined with immersive presence, addiction risk could rise sharply. The metaverse may also enable new digital forms addictions previously only found in the physical world, like exercise and sexual addiction. Even without clinical addiction, the metaverse's appeal may lead many to spend excessive time there, neglecting real-life responsibilities and relationships.

A major concern is the use of addiction by design in the metaverse, where immersive environments and embodied presence may make digital experiences more addictive than those on the internet. With the entire metaverse functioning as an interactive interface, dark patterns—like gamification, sensory effects, and social rewards—can be embedded anywhere. These can be personalized to exploit user vulnerabilities. Given the serious risks, strict limits on such design practices are essential.

5.2.3 The Metaverse and Social Well-Being

Individual well-being depends significantly on social relationships. As social beings, humans tend to lead less fulfilling lives without social interaction. Social relationships can be personal or impersonal. Personal relationships—such as friendships, family ties, and romantic bonds—are built on emotional connection, trust, and mutual support. Impersonal relationships, like professional or economic ones, are goal-oriented and lack emotional closeness. This section focuses on personal relationships, as they are a key condition for well-being.

The rise of the internet has had mixed effects on personal relationships. On the positive side, it offers new ways to stay in touch over distances, helps form new friendships and romantic connections, and can support social learning. However, it may also weaken offline relationships through replacement, displacement, or degradation. For example, excessive internet use can reduce time spent with loved ones or alter mood and behavior in ways that harm intimacy and trust. This concern is captured in the "social displacement hypothesis," which suggests that online interactions may reduce face-to-face socializing (Kraut et al. 1998; Gapsiso and Wilson 2015). Evidence for this is mixed: some studies show declines in in-person interactions, but others find no clear link with increased social media use. Many people use social media to maintain existing personal ties, particularly friendships, especially when face-to-face contact is not possible. However, social media may be less effective in supporting family relationships and could negatively impact family life when those relationships are already strained.

The metaverse is likely to significantly impact personal relationships, particularly *friendships*. Like social media, it enables contact between distant friends, but with the added benefit of embodied interaction. This form of communication closely resembles face-to-face encounters, allowing for richer exchanges through facial expressions, body language, and shared virtual activities—such as concerts or nature explorations—which can strengthen bonds. As a result, the perceived need for in-person meetings may decline. However, heavy metaverse use could reduce face-to-face interactions and negatively affect relationships within one's immediate physical environment, especially with family and partners. This risk increases when VR gear isolates users from their surroundings, unlike smartphones or tablets that still allow for some multitasking.

The metaverse may also encourage the development of purely online friendships, something less common on social media. These immersive connections, while meaningful, could displace time spent on offline relationships. Additionally, AI-powered bots programmed to act as friends or companions may become more common. While these bots might offer emotional support or learning opportunities, they also risk undermining real-world relationships by setting unrealistic expectations or replacing human interaction.

The metaverse can strengthen *family bonds*, especially across long geographical distances, by offering a shared sense of presence and meaningful joint activities, such as virtual meals or exploring digital environments together. Even co-located family members can use it to deepen connections through shared experiences. However, excessive or individual use may harm relationships, creating a displacement effect and leading to parallel lives. To maintain strong and healthy ties, families should either set clear limits on individual metaverse use or make a conscious effort to engage together in the same virtual spaces.

The metaverse has the potential to both enhance and challenge *romantic relationships*. As with friendships and family ties, it can support long-distance relationships by enabling immersive, embodied interactions that help couples feel more connected. In addition, shared activities in virtual environments can deepen emotional bonds. However, excessive individual use of the metaverse for personal interests may reduce time spent together and lead to emotional neglect. A major risk is the possibility of virtual affairs. The metaverse's immersive nature allows users to engage in romantic or sexual activities with others in realistic, embodied ways—without the logistical barriers of physical affairs. These encounters can happen anonymously, making infidelity more accessible and, for some, more tempting.

For existing couples, the metaverse could provide positive experiences through roleplay, fostering intimacy and novelty. However, roleplay also raises ethical concerns—such as deception, the use of child avatars, or unauthorized depictions of real people. The rise of AI-powered "love bots" and "sex bots" in the metaverse poses further complexity. Love bots simulate emotional relationships, while sex bots focus on sexual interaction, often enhanced through VR and haptic technology. These bots may be used for education, therapy, or companionship by those without access to human relationships. Yet, they also present risks to real-life partnerships. Easily formed and concealed, such virtual relationships could be perceived as infidelity by partners and may strain trust and intimacy in existing romantic bonds.

The metaverse could also transform dating. Beyond current internet apps that focus on messaging and profile browsing, users can meet and date in fully simulated environments. They may use dating apps to find matches and then meet in virtual spaces, or encounter potential partners in social hubs within the metaverse. Unlike traditional online dating, metaverse dating can involve immersive experiences that closely resemble real-world dates, including shared activities and even VR-mediated intimacy. While this creates a safer space for initial connections, it also heightens concerns about deception. Users can craft entirely fabricated identities through avatars, AI-enabled behaviors and virtual possessions, leading to disillusionment when relationships shift offline. This risk may call for new forms of identity verification.

5.3 The Metaverse, Politics, and Civil Society

In this section, we explore whether the metaverse is likely to contribute to the social good—that is, whether it can benefit society collectively rather than serving only individual or private interests. The social good encompasses conditions, institutions, resources, or outcomes that enhance the well-being of society as a whole. To assess this, we will examine the metaverse's potential impact on key social domains such as work, education, healthcare, public discourse, and civic engagement. Since the metaverse may come to function as a social environment in its own right—fulfilling essential societal roles and hosting complex interactions—we also consider a deeper, related question: can the metaverse provide the foundation for a good society? To address this, we focus on two core dimensions of a good society: social quality and democratic governance. These two aspects form the basis for our broader evaluation of the metaverse's potential to serve the social good, which we will return to at the end of the section.

5.3.1 *The Social Quality of the Metaverse*

This section explores the concept of *social quality* in the metaverse, focusing on how its social structures and institutions should be arranged to contribute to the common good and support collective well-being. As virtual worlds become increasingly immersive and multifunctional, with users spending more time in them, social complexity will grow. Social institutions could form, either mirroring or diverging from those in the physical world. We first examine types of social organization that could emerge in the metaverse and then identify standards by which to evaluate their quality.

Sociologists describe societies as systems of social relationships that provide structure and help fulfill human needs. These relationships are shaped by social structures such as norms, roles, networks, and institutions, which guide behavior and define interactions. Culture complements these structures by providing shared beliefs and values. The institution of marriage, for example, illustrates how social structures and cultural values jointly shape roles, interactions, and access to resources.

A society is defined by four elements: complex social relationships, supporting social structures, cultural frameworks, and a resource-providing environment. Can the metaverse instantiate these elements? It can host complex relationships, support cultural expression, and be a space where social norms and institutions could be developed. However, it falls short of being a full society due to its dependence on external legal and institutional systems and its inability to fulfill basic human needs like food, shelter, and physical security. At best, the metaverse can support a quasi-society that resembles a real one in complexity but lacks full autonomy.

We will assume that a quasi-society, like a real society, can be assessed in terms of its social quality. We will carry out an initial assessment using the model of social quality provided by Abbott and Wallace (2012), which includes four dimensions: socio-economic security, social cohesion, social inclusion, and conditions for social empowerment. *Socio-economic security*, to start with, refers to access to resources needed for a dignified life. While the metaverse cannot meet all such needs, it can support economic participation, virtual jobs, and services like education and telehealth. Ensuring fair labor standards in metaverse jobs and providing access to beneficial services can enhance its contribution to this dimension.

Social cohesion involves trust, cooperation, and a sense of belonging. The metaverse can foster this through inclusive public spaces, civic engagement opportunities, and support for diverse communities. Preventing fragmentation and encouraging interaction among groups is essential. *Social inclusion* entails equal participation and access to opportunities. This requires equitable access to the metaverse itself, unbiased design, and active measures to prevent discrimination, as discussed in our section on equality and fairness in the previous chapter. Both tech companies and users have a role in building inclusive communities. *Social empowerment*, finally, means that individuals have the capacity to act and make decisions. This depends on factors such as education, health, access to resources, safety, and participation in governance. The metaverse should be designed and operated to enhance user agency and support these conditions.

In conclusion, while the metaverse cannot be a self-governing society, it can function as a quasi-society. By aligning its design and governance with the four dimensions of social quality, it can become a socially beneficial environment that supports the common good and human flourishing in digital space.

5.3.2 Democracy in the Metaverse

In Chap. 4, we examined centralized and decentralized models of metaverse governance and ownership, and how these relate to the common good. In this section, we build on that discussion by evaluating the strengths and weaknesses of various governance models, with particular attention to the question of whether metaverse governance ought to be democratic. Governance refers to the systems and processes by which entities—such as political institutions, corporations, and digital infrastructures—are directed, managed, and held accountable. The metaverse, like the internet, is a complex sociotechnical system that requires governance by different actors and across multiple levels: government regulation, industry self-regulation, and user or civil society participation. Three key parameters define metaverse governance models: the degree of governmental regulation, the extent of metaverse-wide industry self-regulation, and the involvement of users and civil society stakeholders.

Governance is closely linked to metaverse architecture. Centralized models concentrate control in the hands of a few entities, typically private companies. Decentralized models, particularly blockchain-based ones, distribute control among users, often enhancing democratic participation. The current internet operates largely under centralized governance by tech corporations, with varied regulatory input from governments. Advocates of Web3 hope the metaverse will move toward more democratic, decentralized models.

Democratic governance of the metaverse is desirable because the metaverse may serve functions similar to a society or community. People should not have to forfeit democratic rights when transitioning parts of their lives into virtual environments. Democratic models give users a voice and help protect their interests. However, critics argue that the metaverse will largely be developed by private enterprise, and democratic governance could hinder innovation and conflict with ownership rights. Nonetheless, democratic governance is achievable, even under private ownership. Platform operators could retain control over core infrastructure, while users govern the virtual worlds. Governments and nonprofits could play significant roles in ownership and oversight.

Here are four possible democratic governance models for the metaverse:

1. *Decentralized Governance*: A metaverse with a decentralized architecture and decentralized ownership and governance by users (Shapiro and Talmon 2022). Decisions are made directly by users through blockchain-enabled mechanisms.
2. *Centralized Governance with Public or Nonprofit Ownership*: Governmental or nonprofit bodies own and operate platforms, with democratically chosen boards.
3. *Centralized Governance with Corporate Ownership and Stakeholder Participation*: A for-profit entity controls the platform but incorporates democratic elements. Three versions are: company-controlled platforms that host user-governed virtual worlds; user-owned for-profit cooperatives with elected boards that govern platform and virtual worlds; and corporations adopting stakeholder governance, where elected users help shape both platform policies and broader corporate strategy.

4. *Mixed Governance Models*: Public-private partnerships or multistakeholder councils oversee metaverse development. These bodies combine government, industry, and civil society participation to create binding policies and ensure democratic input.

In Chap. 4, we already considered arguments for and against some of these models. We found that decentralized governance models empower users and reduces concentrated control, but they may lead to complexity, unequal participation, and fragmentation. Governance through public or nonprofit ownership offers stability and commitment to the common good, but it may lack innovation and suffer from bureaucratic inefficiency. Centralized governance models with corporate ownership can allow for some democratic participation, but commercial priorities may at times conflict with democratic values and user interests. Mixed models, finally, may feature advantages of the other three models, but disadvantages as well.

Current trends indicate that the metaverse will likely be dominated by private enterprise, making model 3 the most realistic and attainable form of democratic governance under such conditions. Government regulation remains a key tool for democratizing metaverse governance in this scenario. Democratic governments can impose rules that protect users and promote fairness, such as transparency, ethical practices, and stakeholder rights. Users and civil society groups also have a responsibility here—to organize, raise awareness, and push for policies, design choices and governance models that prioritize public interest.

5.3.3 The Metaverse and the Good of Society

Research indicates that many experts and members of the public believe the internet is not contributing positively to society (Smith and Olmstead 2018; Wike et al. 2022). Common concerns include the spread of misinformation, a decline in civility and constructive public discourse, harmful effects on democracy and trust in institutions, increasing social fragmentation, and negative impacts on mental health. Many now doubt whether these negative consequences are outweighed by the internet's benefits, such as social connection, economic opportunity, and access to information.

The metaverse will likely have many of the same applications as the internet, along with new ones. Will it play a similarly ambivalent role in society? Will its benefits clearly outweigh its drawbacks, or will its negative consequences prove to be more significant over time? It is not possible to provide an uncontroversial answer to these questions, as it depends on how the metaverse will be developed, governed, and used, as well as on broader social, political, and economic developments that extend beyond the metaverse itself. At this stage, what we can meaningfully do is map the most plausible benefits and risks associated with the metaverse, particularly those that relate to the social good.

As discussed in Chap. 3, the metaverse is expected to bring a wide range of societal benefits. Key areas such as education and healthcare stand to gain significantly, alongside several major industries. Sectors likely to benefit most include technology and software development, telecommunications, e-commerce, media, manufacturing, banking, and professional services (Elmasry et al. 2022). These developments are also expected to drive job growth across many of these fields. Beyond economic impacts, the metaverse could support social empowerment and individual well-being by offering new opportunities for work, deeper social connection, immersive learning and skill-building, and creative self-expression. It may also increase social participation for marginalized groups and people with disabilities, while making civic engagement more inclusive through innovations like virtual town halls and participatory digital platforms. Finally, as the next section will explore, the metaverse could contribute to environmental sustainability by shifting some physical activities into virtual space and using simulations to support more sustainable design and planning.

The metaverse also poses serious risks to the common good. As discussed earlier in this chapter, its use can negatively affect mental health and social well-being. As noted in Chap. 4, it also poses serious threats to individual rights, including privacy, security, autonomy, equality, and property. Without strong safeguards, user privacy could be virtually eliminated, leaving individuals vulnerable to manipulation through detailed tracking of their behavior, emotions, and social interactions. Public discourse and democratic processes may be further undermined by deepfakes, immersive propaganda, and algorithmic echo chambers that are even more persuasive than those on today's internet. Corporate control could further erode the public value of virtual spaces by transforming both public and private areas into highly commercialized, surveilled, and restricted environments, where profit outweighs user rights and equitable access. Moreover, increased reliance on virtual socialization could weaken real-world community ties, local cultures, and shared civic life. Finally, the significant resource demands and carbon emissions associated with building and maintaining the metaverse could cause serious environmental harm—potentially outweighing its contributions to sustainability outlined earlier.

Ultimately, the extent to which these benefits and risks materialize will depend on how the metaverse is developed, governed, and used—as well as on broader social, political, and economic developments. At this stage, it is not possible to definitively assess the metaverse's overall contribution to the social good. What is essential, however, is to remain aware of its potential impacts and to take deliberate steps toward shaping its development in ways that maximize its benefits while actively addressing and minimizing its risks.

5.4 The Metaverse and Environmental Sustainability

In Chap. 3, we noted that the metaverse demands an IT infrastructure capable of processing and storing data at scales far beyond current capabilities. Building it would require vast new investments in cloud and edge computing, data centers, and the manufacturing of millions of powerful computers, graphics processors, and immersive devices. This will place heavy demands on energy and resources, raising serious environmental concerns. There will be significant environmental costs associated with building and producing infrastructure, the generation of e-waste from hardware disposal, and the consumption of energy. In this section, we will assess the negative impacts on the environment that could result as well as ways to mitigate these impacts. We will consider potential positive environmental impacts of the metaverse as well.

5.4.1 Negative Environmental Impacts

The metaverse poses several serious environmental risks, including resource depletion for hardware and infrastructure production, the generation of electronic waste, significant water consumption by data centers, high energy use, and increased greenhouse gas emissions. Among these, energy consumption and its associated emissions are expected to be particularly severe. The metaverse will likely be far more energy-intensive than today's internet due to its immersive 3D, multisensory environments, which require vastly greater computing power and data throughput than 2D content. Technologies like virtual and augmented reality depend on high-performance GPUs, rapid 5G networks, and massive data centers. One study found that high-end VR gamers generate about 0.91 metric tons (MT) of CO_2 per year (Dellinger 2020)—a notable figure given that per capita emissions in the Global North range from 4 to 7 MT annually.

Artificial intelligence (AI) will also be central to the metaverse, enabling applications such as computer vision, language processing, and generative content—all of which require immense data processing and energy. Digital twins, which replicate physical systems in virtual form, are expected to be widely deployed and are particularly energy-intensive, relying on the Internet of Things (IoT) and potentially millions of sensors to collect real-time data (Zhang et al. 2022). Blockchain technology, another key component, is also energy-demanding: a single bitcoin transaction consumes 2148 KWh—equivalent to 160,000 credit card transactions or 73 days of electricity use by an average U.S. household (Biswas et al. 2023). While more sustainable blockchain solutions are emerging, overall energy demands remain a major concern.

The vast increase in computing capacity required for the metaverse could significantly raise the IT sector's contribution to global emissions, even with ongoing advances in semiconductor miniaturization. At present, global IT infrastructure already accounts for at least 1.8–2.8% of worldwide emissions (Freitag et al. 2021), and the rise of metaverse technologies could easily push this figure beyond 3%.

5.4.2 Greening the Metaverse

Researchers and developers have begun addressing the metaverse's environmental costs by proposing solutions grounded in energy efficiency, sustainability, and circular economy principles. Zhang et al. (2022) review methods for greening the metaverse, arguing that VR and augmented reality (AR) can become significantly more energy efficient through software and algorithmic improvements. They also review energy-saving techniques for AI, blockchain, and digital twins, focusing on optimized algorithms, machine learning models, communication networks, data centers, and sensors.

In the AI sector, energy efficiency is gaining attention. Beyond using AI to boost efficiency in other fields, efforts are being made to reduce AI's own energy use (Bolón-Canedo et al. 2024). Schwartz et al. (2020) advocate for balancing accuracy with computational efficiency by reporting floating-point operations (FPOs) as a metric. Developers should aim to minimize FPOs while maintaining performance. Similarly, the blockchain industry is adopting more energy-efficient consensus mechanisms, moving from Proof-of-Work to Proof-of-Stake and other alternatives (Fernández-Caramés and Fraga-Lamas 2024).

The ecological impact of data centers, crucial to the metaverse, is also under scrutiny. Pennington et al. (2023) propose a four-part strategy for green data centers: using renewable energy, enhancing efficiency, applying circular economy practices to infrastructure, and reducing water usage. Cao et al. (2022) support similar strategies and recommend reusing waste heat and implementing carbon market tools like cap-and-trade and carbon taxes. The challenge is to green data centers quickly enough to offset their growth driven by the metaverse and other digital trends.

These combined efforts could reduce the metaverse's ecological footprint and that of digital infrastructure more broadly. However, even with improved efficiency, the metaverse will still consume large amounts of energy and produce negative environmental effects through construction, manufacturing, and e-waste. It has been claimed, however, that the metaverse will also have positive environmental impacts. We will now proceed to examine these claims.

5.4.3 Environmental Savings Through Virtual Replacements

As discussed in Chap. 3, the metaverse can support a wide range of activities, including commerce, education, office work, leisure, certain healthcare services, and parts of industrial processes. Many of these activities will replace physical counterparts. For example, virtual meetings may substitute for in-person ones. In this way, the metaverse can contribute to *dematerialization*, which refers to reducing the material and energy intensity of goods and services through digitization. Substituting physical practices with virtual ones could lead to significant environmental benefits (Kshetri and Dwivedi 2023; Stoll et al. 2022), as virtual alternatives typically require less infrastructure, less transportation, and fewer physical goods.

We have already observed this substitution effect with the current internet. The internet supports many traditionally physical practices—shopping, banking, education, media consumption, and social interaction. Some of these have clearly declined in the physical realm: people shop less in stores, visit banks and government offices less often, go to cinemas less frequently, and consume fewer printed newspapers. The internet has therefore already contributed to a process of digital substitution and dematerialization, and this trend may continue with the metaverse.

The metaverse may reduce the need for physical practices and products in specific ways. One area is commercial and public buildings. As work, education, commerce, and entertainment shift partly to the metaverse, the demand for office space, educational institutions, retail stores, event venues, and government buildings may decline. Residential real estate is unlikely to be affected to the same extent, as homes serve essential physical needs.

A decrease in demand for physical goods is also likely. The internet already reduced the need for items such as physical media, physical money, physical games, and physical media creation tools. In addition, the emergence of the metaverse could result in reduced demand for physical clothing, jewelry and makeup because physical encounters are partially replaced by encounters between avatars. Reductions in commercial and governmental buildings could also result in less need for building materials, furniture, decorations, office equipment, travel gear and other items needed in these spaces or for travel to them.

Third, a reduction in the need for transportation is also likely. As more people work, learn, shop, and socialize in the metaverse, less physical travel beyond the home may be necessary, lowering reliance on public and private transport and associated infrastructure. In addition, reduced construction and manufacture due to the reduced need for commercial and public venues and for physical goods could also result in less freight transportation.

Zhao and You (2023) analyzed the potential environmental impact of the metaverse in the U.S. by 2050, assuming large-scale adoption. They examined five key use cases—work, travel, education, NFTs, and gaming— and found that metaverse growth in these areas would accelerate decarbonization and improve air quality. They estimated a 10% reduction in national energy use and a 10–23% decrease in air pollutants by 2050. The largest energy savings would come from virtual workplaces (3337 Petajoule), followed by travel (2050 PJ) and education (1348 PJ).

However, the assumption that digitization will reduce physical goods and practices, thereby producing environmental benefits, has been questioned. Similar predictions were made during the early days of the internet, when many expected it to reduce travel, transportation, and physical production. Yet, global traffic, freight transport, construction, and industrial output have all increased substantially. Even if digitization has contributed to some dematerialization, the fact is that an increase in environmental impact has occurred.

5.4 The Metaverse and Environmental Sustainability

There are strong indications that digitization has not led to widespread dematerialization, except in a few specific areas. One key reason is the emergence of *rebound effects*—situations where the expected energy or resource savings from more efficient technologies are offset, or even surpassed, by increased consumption. These effects often occur because greater efficiency lowers costs for users, increasing their spending power. As a result, people may use the improved technology more intensively or redirect their savings toward other products and services, ultimately undermining the environmental benefits of the efficiency gains.

Greening et al. (2000) outline three types of rebound effects. A *direct rebound effect* occurs when savings from a more efficient technology lead to increased use of that same technology. An *indirect rebound effect* happens when cost savings are spent on other goods or services. Finally, an *economy-wide rebound effect* is a macro-effect in which improvements in technology lead to increased production and economic growth, driving higher overall resource use.

Gossart (2015) argues that such effects have been especially strong in information and communication technologies (ICT). He notes that miniaturization has reduced material and energy demands per device, but this has led to more powerful devices being used more widely and frequently. Studies cited by Gossart show that cost savings in ICT use can lead to higher resource use elsewhere. For example, the rise of telework has enabled some people to live farther from their workplaces, increasing their commuting distances. Likewise, savings from teleshopping, telecommuting, and teleconferencing may be offset by increased travel for other purposes. On a larger scale, the internet has been linked to the emergence of a global, networked economy that enables businesses to streamline operations and expand their reach, ultimately driving further growth in production and transportation (Castells 2002).

It is likely that the metaverse will generate similar rebound effects. Direct rebound could occur as people consume more virtual services simply because they are cheaper and easier to access. For instance, low-cost virtual concerts might lead people to attend more performances than they would physically. Similarly, users might engage more frequently with virtual shopping or social events because they are more convenient and less costly than physical alternatives. Savings from digitization may also result in indirect rebound effects, as lower costs in one area increase consumption in others. A user who saves on commuting or travel costs by working in the metaverse might spend those savings on more energy-intensive leisure activities or travel.

Economy-wide rebound effects may result as well. As we saw in Chap. 3, the business case for the metaverse includes not only gains from dematerialization and efficiency but also its potential to stimulate the physical economy. Virtual experiences could increase demand for real-world goods and services. For example, virtual tourism could lead to an increased demand for physical tourism, as consumers discover beautiful remote places that they would like to visit in person. Similarly, virtual shopping could expose consumers to new products, which could lead to increased consumption. By strengthening the networked economy, the metaverse may also intensify globalization, leading to increased freight transport and business travel.

Whether these rebound effects will occur—and to what extent—depends on broader economic, social, and policy conditions. They are not inherent to the technologies themselves. As Font Vivanco et al. (2016) and Freire-González and Ho (2022) argue, rebound effects can be mitigated with targeted policy interventions. These might include resource and energy efficiency standards, environmental taxes, emissions trading systems, or lifestyle-oriented policies designed to shift consumer behavior. With the right policy framework in place, it may be possible to limit rebound effects and ensure that the metaverse contributes to net environmental gains through dematerialization. However, this could be challenging under current political and economic systems, which tend to prioritize growth over sustainability.

5.4.4 Environmental Savings Through Environmental Simulations

Metaverse simulations can be used to promote more sustainable products and practices in the physical world. Most importantly, this can be achieved through digital twins (Rojek et al. 2021; Verzelen et al. 2021; Yang et al. 2022). Digital twins are used for the environmental design, monitoring, and optimization of various systems, such as buildings, machines, industrial processes, energy and transportation systems, cities, and natural ecosystems. These twins include sustainability metrics to monitor and optimize environmental performance. Some are dedicated entirely to sustainability—so-called digital twins for sustainability—while others incorporate environmental metrics alongside broader functions. Typically, simulations are run to analyze different scenarios involving various configurations or interventions, and sustainability metrics help determine which scenario yields the best outcomes in terms of energy use, emissions, and resource consumption. Action can then be taken in the physical world to implement the most sustainable scenarios.

Studies suggest that the environmental savings from digital twins can be substantial. An Accenture report claims that energy use in commercial and residential buildings—responsible for about one-third of global emissions—could be reduced by 50–80% through digital twins (Verzelen et al. 2021). It also describes a case where a digital twin optimized a pharmaceutical manufacturing plant, reducing energy use and CO_2 emissions by 80% per year, water usage by 91%, chemical use by 94%, and cutting 321 tons of waste annually. Another report presents a case from Nanyang Technological University, where digital twins led to more than 40% energy savings and a 17.8-kiloton reduction in carbon emissions (Ottinger et al. 2021).

Environmental digital twins are a special category focused on modeling natural ecosystems using data from sensors, satellite imagery, climate models, and other environmental tools (Siddorn et al. 2022). These dynamic simulations support climate change modeling, ecosystem monitoring and conservation, water management, urban planning, and agriculture. While primarily used for research, planning, and development, the insights they provide could also be used in ways that are not environmentally sustainable.

Another potential contribution of the metaverse to sustainability is through immersive simulations for environmental education and awareness (Cho and Park 2023; Kleinlogel et al. 2023). While digital twins can support learning, other simulations can also provide rich educational experiences. These VR experiences may allow users to interact with entire ecosystems over compressed time spans or explore microscopic environments. They can simulate environmental challenges and responses or recreate natural ecosystems with lifelike flora and fauna. They can also teach and reinforce sustainable behaviors. In this way, the metaverse can offer powerful, immersive tools to foster environmental understanding and awareness.

5.5 Conclusion

In this chapter, we explored the potential implications of the metaverse for well-being, for the good of society, and for environmental sustainability. We found that the metaverse can have both strongly positive and strongly negative implications for well-being, depending on how it is developed and used. It was considered how aspects of well-being such as physical and mental health, pleasure, life satisfaction, meaningfulness and personal growth can be harmed or benefited. Mental health was found to be a major risk factor, due to the dangers such as metaverse addiction and overuse, social withdrawal, and social comparison and low self-esteem. As for social well-being, it was found that although friendships, family bonds and romantic relationships can flourish in the metaverse, there are also major risk factors such as virtual affairs, friendship bots and love bots. Developers and operators of metaverse system have a responsibility to consider the consequences for well-being of different metaverse designs and operations. This includes a careful consideration of new services and features for their potential effects on well-being, including mental health and social relations.

We examined the quality of social systems in the metaverse and its potential contribution to the social good. While the metaverse may not support a fully self-governing society, it can function as a kind of virtual quasi-society with key features of real-world social systems. The quality of this social system will depend on its support for socio-economic security, inclusion, cohesion, and empowerment. We also explored whether metaverse governance should be democratic, arguing that the metaverse's pervasive role in users' lives justifies giving them influence over its design and operations. Various democratic models were discussed with different ownership structures. Finally, we considered both benefits and risks of the metaverse for society: for example, benefits such as expanded access to education and civic participation, and risks such as surveillance, corporate control, and social fragmentation. Ultimately, whether the metaverse serves the social good will depend on how it is shaped—through governance choices, design decisions, and broader societal developments.

At last, we looked at the metaverse's environmental impact. While development and use may lead to significant environmental harm, there are also opportunities for environmental benefits—especially through substitution effects and the use of digital twins. However, we cautioned that rebound effects could offset these gains unless addressed by targeted policy. In conclusion, the metaverse has the potential for both positive and negative environmental outcomes, depending on how it is developed and used.

References

Abbott, Pamela, and Claire Wallace. 2012. Social Quality: A Way to Measure the Quality of Society. *Social Indicators Research* 108 (1): 153–167.

Araigy, M. (2018). Instagram Usage and Its Relation to Self Esteem among Lebanese Young Adults. *International Journal of Humanities and Social Science* 8 (9). https://doi.org/10.30845/ijhss.v8n9p12.

Bányai, F., Á. Zsila, O. Király, A. Maraz, Z. Elekes, M. D. Griffiths, C. S. Andreassen, and Z. Demetrovics. 2017. Problematic Social Media Use: Results from a Large-Scale Nationally Representative Adolescent Sample. *PloS one* 12 (1): e0169839. https://doi.org/10.1371/journal.pone.0169839.

Barreda-Ángeles, M., and T. Hartmann. 2022. Hooked on the Metaverse? Exploring the Prevalence of Addiction to Virtual Reality Applications. *Frontiers in Virtual Reality* 3:1031697. https://doi.org/10.3389/frvir.2022.1031697.

Biswas, D., H. Jalali, A. H. Ansaripoor, and P. De Giovanni. 2023. Traceability vs. Sustainability in Supply Chains: The Implications of Blockchain. *European Journal of Operational Research* 305 (1): 128–147. https://doi.org/10.1016/j.ejor.2022.05.034.

Bolón-Canedo, V., M. Rey-Area, P. Tino, and A. Alonso-Betanzos. 2024. A Review of Green Artificial Intelligence: Towards a More Sustainable Future. *Neurocomputing* 565:128096. https://doi.org/10.1016/j.neucom.2024.128096.

Brey, Philip. 2012. Well-Being in Philosophy, Psychology, and Economics. In *The Good Life in a Technological Age*, ed. Philip Brey, Adam Briggle, and Edward Spence, 15–34. London: Routledge.

Camilleri, C., J. T. Perry, and S. Sammut. 2021. Compulsive Internet Pornography Use and Mental Health: A Cross-Sectional Study in a Sample of University Students in the United States. *Frontiers in Psychology* 11:613244. https://www.frontiersin.org/articles/10.3389/fpsyg.2020.613244.

Cao, Zhiwei, Xin Zhou, Han Hu, Zhi Wang, and Yonggang Wen. 2022. Towards a Systematic Survey for Carbon Neutral Data Centers. *arXiv*: 2110.09284. http://arxiv.org/abs/2110.09284. Accessed 29 Mar 2025.

Castells, Manuel. 2002. *The Internet Galaxy: Reflections on the Internet, Business, and Society*. Oxford: Oxford University Press.

Cho, Y., and K. S. Park. 2023. Designing Immersive Virtual Reality Simulation for Environmental Science Education. *Electronics* 12 (2): 315. https://doi.org/10.3390/electronics12020315.

Cingel, D. P., M. C. Carter, and H.-V. Krause. 2022. Social Media and Self-Esteem. *Current Opinion in Psychology* 45:101304. https://doi.org/10.1016/j.copsyc.2022.101304.

De Alarcón, R., J. I. de la Iglesia, N. M. Casado, and A. L. Montejo. 2019. Online Porn Addiction: What We Know and What We Don't—A Systematic Review. *Journal of Clinical Medicine* 8 (1): 91. https://doi.org/10.3390/jcm8010091.

Dellinger, Aj 2020. The Unseen Environmental Cost of Gaming. Mic, January 29. https://www.mic.com/impact/gamings-environmental-impact-is-bigger-than-you-think-21753800. Accessed 29 Mar 2025.

References

Elmasry, Tarek, Eric Hazan, Hamza Khan, Greg Kelly, Shivam Srivastava, Lareina Yee, and Rodney W. Zemmel. 2022. *Value Creation in the Metaverse: The Real Business of the Virtual World*. McKinsey & Company. https://www.mckinsey.com/~/media/mckinsey/business%20 functions/marketing%20and%20sales/our%20insights/value%20creation%20in%20the%20 metaverse/Value-creation-in-the-metaverse.pdf. Accessed 1 Apr 2025.

Elsey, J. W. B., K. van Andel, R. B. Kater, I. M. Reints, and M. Spiering. 2019. The Impact of Virtual Reality Versus 2D Pornography on Sexual Arousal and Presence. *Computers in Human Behavior* 97:35–43. https://doi.org/10.1016/j.chb.2019.02.031.

Fernández-Caramés, T.M., and P. Fraga-Lamas. 2024. A Comprehensive Survey on Green Blockchain: Developing the Next Generation of Energy Efficient and Sustainable Blockchain Systems. *arXiv*: 2410.20581. https://doi.org/10.48550/arXiv.2410.20581. Accessed 29 Mar 2025.

Floridi, L. 2022. Metaverse: A Matter of Experience. *Philosophy & Technology* 35 (3): 73. https://doi.org/10.1007/s13347-022-00568-6.

Font Vivanco, D., R. Kemp, and E. van der Voet. 2016. How to Deal with the Rebound Effect? A Policy-Oriented Approach. *Energy Policy* 94:114–125. https://doi.org/10.1016/j.enpol.2016.03.054.

Forsman, A. K., and J. Nordmyr. 2017. Psychosocial Links Between Internet Use and Mental Health in Later Life: A Systematic Review of Quantitative and Qualitative Evidence. *Journal of Applied Gerontology* 36 (12): 1471–1518. https://doi.org/10.1177/0733464815595509.

Freire-González, J., and M. S. Ho. 2022. Policy Strategies to Tackle Rebound Effects: A Comparative Analysis. *Ecological Economics* 193:107332. https://doi.org/10.1016/j.ecolecon.2021.107332.

Freitag, C., M. Berners-Lee, K. Widdicks, B. Knowles, G. S. Blair, and A. Friday. 2021. The Real Climate and Transformative Impact of ICT: A Critique of Estimates, Trends, and Regulations. *Patterns* 2 (9): 100340. https://doi.org/10.1016/j.patter.2021.100340.

Gapsiso, Nuhu Diraso, and Joseph Wilson. 2015. The Impact of the Internet on Teenagers' Face-to-Face Communication. *Journal of Studies in Social Sciences* 13 (2):

Gossart, Cédric. 2015. Rebound Effects and ICT: A Review of the Literature. In *ICT Innovations for Sustainability*, ed. Lorenz M. Hilty and Bernard Aebischer, 435–448. Cham: Springer International Publishing.

Greening, L. A., D. L. Greene, and C. Difiglio. 2000. Energy Efficiency and Consumption—The Rebound Effect—A Survey. *Energy Policy* 28 (6): 389–401. https://doi.org/10.1016/S0301-4215(00)00021-5.

Kleinlogel, E. P., M. Schmid Mast, L. A. Renier, M. Bachmann, and T. Brosch. 2023. Immersive Virtual Reality Helps to Promote Pro-Environmental Norms, Attitudes and Behavioural Strategies. *Cleaner and Responsible Consumption* 8:100105. https://doi.org/10.1016/j.clrc.2023.100105.

Kraut, R., M. Patterson, V. Lundmark, S. Kiesler, T. Mukophadhyay, and W. Scherlis. 1998. Internet Paradox: A Social Technology that Reduces Social Involvement and Psychological Well-Being? *American Psychologist* 53:1017–1031. https://doi.org/10.1037/0003-066X.53.9.1017.

Kshetri, N., and Y. K. Dwivedi. 2023. Pollution-Reducing and Pollution-Generating Effects of the Metaverse. *International Journal of Information Management* 69:102620. https://doi.org/10.1016/j.ijinfomgt.2023.102620.

Lockhart, Marisa. 2019. *The Relationship Between Instagram Usage, Content Exposure, and Reported Self-Esteem*. George Mason University.

Nutt, D. J., A. Lingford-Hughes, D. Erritzoe, and P. R. A. Stokes. 2015. The Dopamine Theory of Addiction: 40 years of Highs and Lows. *Nature Reviews Neuroscience* 16 (5): 305–312. https://doi.org/10.1038/nrn3939.

Orben, A. 2020. Teenagers, Screens and Social Media: A Narrative Review of Reviews and Key Studies. *Social Psychiatry and Psychiatric Epidemiology* 55 (4): 407–414. https://doi.org/10.1007/s00127-019-01825-4.

Ottinger, Nipun Bajaj, Erica Jordan Stein, Mark Gibson Crandon, and Akanksha Jain. 2021. Digital Twin: The Age of Aquarius in Construction and Real Estate. Ernst & Young Global

Limited. https://www.scl.org.au/sites/default/files/Digital%20twin%20-%20the%20Age%20of%20Aquarius%20in%20construction%20and%20real%20estate.pdf. Accessed 1 Apr 2025

Paakkari, L., J. Tynjälä, H. Lahti, K. Ojala, and N. Lyyra. 2021. Problematic Social Media Use and Health among Adolescents. *International Journal of Environmental Research and Public Health 18* (4): 1885. https://doi.org/10.3390/ijerph18041885.

Pennington, James, Marc Walsh, Aisling Curtin, and Rebecca Murphy. 2023. *Green Data—A 4 Pillar Approach for Green Data Centres*. Deloitte. https://www2.deloitte.com/content/dam/Deloitte/uk/Documents/technology-media-telecommunications/deloitte-uk-green-data-2023.pdf. Accessed 29 Mar 2025

Reed, Nelson, and Katie Joseff. 2022. *Kids and the Metaverse: What Parents, Policymakers, and Companies need to Know*. Common Sense. https://www.commonsensemedia.org/sites/default/files/featured-content/files/metaverse-white-paper-1.pdf. Accessed 29 Mar 2025.

Robbins, T. W., and L. Clark. 2015. Behavioral addictions. *Current Opinion in Neurobiology* 30:66–72. https://doi.org/10.1016/j.conb.2014.09.005.

Rojek, I., D. Mikołajewski, and E. Dostatni. 2021. Digital Twins in Product Lifecycle for Sustainability in Manufacturing and Maintenance. *Applied Sciences* 11 (1): 31. https://doi.org/10.3390/app11010031.

Ryding, F. C., and D. J. Kuss. 2020. The Use of Social Networking Sites, Body Image Dissatisfaction, and Body Dysmorphic Disorder: A Systematic Review of Psychological Research. *Psychology of Popular Media* 9 (4): 412–435. https://doi.org/10.1037/ppm0000264.

Saiphoo, A. N., and Z. Vahedi. 2019. A Meta-Analytic Review of the Relationship Between Social Media Use and Body Image Disturbance. *Computers in Human Behavior* 101:259–275. https://doi.org/10.1016/j.chb.2019.07.028.

Schüll, Natasha Dow. 2014. *Addiction by Design: Machine Gambling in Las Vegas*. Princeton, NJ: Princeton University Press.

Schwartz, R., J. Dodge, N. A. Smith, and O. Etzioni. 2020. Green AI. *Communications of the ACM* 63 (12): 54–63. https://doi.org/10.1145/3381831.

Seligman, Martin E. 2011. *Flourish: A Visionary New Understanding of Happiness and Well-Being*. New York: Simon and Schuster.

Shapiro, Ehud, and Nimrod Talmon. 2022. Foundations for Grassroots Democratic Metaverse. *arXiv*. https://doi.org/10.48550/arXiv.2203.04090. Accessed 29 Mar 2025

Siddorn, J., G. Blair, D. Boot, J. Buck, A. Kingdon, A. Kloker, A. Kokkinaki, G. Moncoiffe, E. Blyth, M. Fry, R. Heaven, E. Lewis, B. Marchant, B. Napier, C. Pascoe, J. Passmore, S. Pepler, P. Townsend, and J. Watkins. 2022. *An Information Management Framework for Environmental Digital Twins (IMFe)*. National Oceanography Centre. https://doi.org/10.5281/zenodo.7004351. Accessed 1 Apr 2025.

Smith, Aaron, and Kenneth Olmstead. 2018. *Declining Majority of Online Adults Say the Internet Has Been Good for Society*. Pew Research Center. April 30. https://www.pewresearch.org/internet/wp-content/uploads/sites/9/2018/04/PI_2018.04.30_Internet-Good-Bad_FINAL.pdf. Accessed 1 Apr 2025.

Sniewski, L., P. Farvid, and P. Carter. 2018. The Assessment and Treatment of Adult Heterosexual Men with Self-Perceived Problematic Pornography Use: A Review. *Addictive Behaviors* 77:217–224. https://doi.org/10.1016/j.addbeh.2017.10.010.

Stavropoulos, Vasileios, Tyrone L. Burleigh, Charlotte L. Beard, Rapson Gomez, and Mark D. Griffiths. 2019. Being There: A Preliminary Study Examining the Role of Presence in Internet Gaming Disorder. *International Journal of Mental Health and Addiction* 17:880–890.

Sternlicht, Lin, and Aaon Sternlicht. 2022. A New Age of Digital Addiction—What the Metaverse Means for Mental Health and Digital Addiction. https://www.familyaddictionspecialist.com/blog/a-new-age-of-digital-addiction-what-the-metaverse-means-for-mental-health-and-digital-addiction. Accessed 29 Mar 2025.

Stevens, M. W., D. Dorstyn, P. H. Delfabbro, and D. L. King. 2021. Global Prevalence of Gaming Disorder: A Systematic Review and Meta-Analysis. *The Australian and New Zealand Journal of Psychiatry* 55 (6): 553–568. https://doi.org/10.1177/0004867420962851.

References

Stoll, C., U. Gallersdörfer, and L. Klaaßen. 2022. Climate Impacts of the Metaverse. *Joule* 6 (12): 2668–2673. https://doi.org/10.1016/j.joule.2022.10.013.

Sun, Y., and Y. Zhang. 2021. A Review of Theories and Models Applied in Studies of Social Media Addiction and Implications for Future Research. *Addictive Behaviors* 114:106699. https://doi.org/10.1016/j.addbeh.2020.106699.

Valkenburg, P. 2021. Social Media Use and Well-Being: What We Know and What We Need to Know. *Current Opinion in Psychology* 45:101294. https://doi.org/10.1016/j.copsyc.2021.12.006.

Vandenbosch, L., J. Fardouly, and M. Tiggemann. 2022. Social Media and Body Image: Recent Trends and Future Directions. *Current Opinion in Psychology* 45:101289. https://doi.org/10.1016/j.copsyc.2021.12.002.

Verzelen, Florence, Peter Lacy, and Nigel Stacey. 2021. Designing Disruption: The Critical Role of Virtual Twins in Accelerating Sustainability. Accenture. https://www.accenture.com/content/dam/accenture/final/a-com-migration/r3-3/pdf/pdf-147/accenture-virtual-twin-and-sustainability.pdf. Accessed 1 Apr 2025.

Webster, A. 2023, March 23. *Tim Sweeney Explains How the Metaverse Might Actually Work.* The Verge. https://www.theverge.com/2023/3/23/23652928/tim-sweeney-interview-epic-games-fortnite-metaverse

Wike, Richard, Laura Silver, Janell Fetterolf, Christine Huang, Sarah Austin, Laura Clancy, and Sneha Gubbala. 2022. *Social Media Seen as Mostly Good for Democracy Across Many Nations, But U.S. Is a Major Outlier*. Pew Research Center. December 6. https://www.pewresearch.org/wp-content/uploads/sites/20/2022/12/PG_2022.12.06_Online-Civic-Engagement_REPORT.pdf. Accessed 1 Apr 2025.

Yang, B., Z. Lv, and F. Wang. 2022. Digital Twins for Intelligent Green Buildings. *Buildings* 12 (6): 856. https://doi.org/10.3390/buildings12060856.

Zhang, S., W.Y.B. Lim, W.C. Ng, Z. Xiong, D. Niyato, X.S. Shen, and C. Miao. 2022. Towards Green Metaverse Networking Technologies, Advancements and Future Directions. *arXiv*. https://doi.org/10.48550/arXiv.2211.03057. Accessed 29 Mar 2025.

Zhao, N., and F. You. 2023. The Growing Metaverse Sector Can Reduce Greenhouse Gas Emissions by 10 Gt CO_2e in the United States by 2050. *Energy & Environmental Science* 16 (6): 2382–2397. https://doi.org/10.1039/D3EE00081H.

Chapter 6
How Can the Metaverse be Developed Responsibly?

> *"We can't build a better Internet unless we learn from past mistakes, innovate responsibly, and join forces together to create an Open, Safe, and Inclusive Metaverse."—X Reality Safety Intelligence (XRSI)*

Abstract This chapter of *The Metaverse: A Critical Assessment* explores how the metaverse can be developed and governed responsibly. Responsible development and governance involve not only maximizing potential benefits but also proactively mitigating risks by upholding rights, maintaining ethical standards, and promoting individual and collective well-being—ensuring that the metaverse serves the public good. It also requires meaningful engagement from all stakeholders involved in its design, deployment, and use. To support this, the chapter presents strategies, guidelines, and methods for addressing social and ethical challenges across the metaverse's lifecycle. Key questions include: How can social and ethical risks be effectively identified and mitigated? What roles and responsibilities do different actors hold, and how can they collaborate toward shared goals? The chapter begins with a general multistakeholder framework for responsible governance, followed by approaches to responsible development and operation. Each section considers the actions key actors can take—and the responsibilities they bear—to ensure the metaverse evolves in a just, inclusive, and accountable way.

Keywords Metaverse · Responsible innovation · Governance · Ethics · Stakeholders · Design and development · Platform operation · Ethics guidelines

6.1 Introduction

In the preceding chapters of this book, we delved into the concept of the metaverse, examining its foundations, potential benefits, and the various risks it presents. In this chapter, we turn our attention to how the metaverse can be developed and governed responsibly, taking into account the opportunities and challenges previously outlined. Responsible development entails not only maximizing the potential benefits but also proactively addressing and mitigating the associated risks. This requires a strong commitment to upholding rights and ethical standards within virtual environments, promoting individual and collective well-being, and ensuring that the evolution of the metaverse serves the broader interests of society.

In this chapter, our focus will be strategies, guidelines, and methods for addressing social and ethical issues in the development, deployment, and use of the metaverse. The questions we will try to answer are: How can social and ethical issues in the metaverse be successfully addressed and mitigated? What are good strategies, methods, and best practices for doing so? What are the roles and responsibilities of different actors in this process, and how can they collaborate to work toward joint outcomes?

We will proceed as follows. In the next section, Sect. 6.2, we will present a general, multistakeholder approach for responsible governance of the metaverse. In the sections that follow, we will then focus on responsible governance of two specific practices: responsible governance of metaverse development (Sect. 6.3), and responsible governance of metaverse operations (Sect. 6.4). We will consider, in both sections, what key actors in these processes can and should do in order to act responsibly. In an annex, a proposal is made for ethical guidelines for the responsible development, deployment, and use of the metaverse. Such guidelines provide guidance for processes of responsible governance as discussed in this chapter.

6.2 Metaverse Ethics and Multistakeholder Governance

Responsible development, management, and use of the metaverse requires a model of responsible governance. As defined in the previous chapter, governance refers to the systems and processes by which entities—such as political institutions, corporations, and digital infrastructures—are directed, managed, and held accountable. The goal of responsible governance is not merely to identify risks or challenges, but to guide stakeholders in making decisions that promote well-being and serve the broader public interest. This also applies to the metaverse, where responsible development, management, and use should aim to prevent harm, support core values, and generate positive outcomes.

To attain these outcomes, there is a need for various actors to act responsibly. No single stakeholder can ensure a beneficial, ethical, and just metaverse that protects values such as freedom, privacy, and equality. Instead, outcomes are shaped through multistakeholder governance processes that distribute responsibilities and guide action. We refer to this model as responsible or ethical metaverse governance.

6.2.1 Principles of Responsible Governance

How can such a governance approach be realized? In this book, we can make proposals for it. However, ultimately, it is the stakeholders themselves who will have to develop it. They will have to negotiate who is responsible for what and how they will collaborate in realizing ethical practices with respect to the metaverse. Our task here is to outline the foundations and suggest which responsibilities should be allocated, setting the stage for more concrete developments.

Governance requires a governed entity (the metaverse) and a governing body—a collection of actors who influence its direction. For the metaverse, relevant actors include private companies, trade and standards organizations, governments, regulators, international institutions, civil society groups, and advocacy networks. Clearly, the governing body is not a cohesive unit but a loose group of interrelated actors, which can often raise issues of responsibility and jurisdiction.

To better understand governance, we can distinguish between two overlapping processes: *development* and *deployment*. Development refers to the design and creation of metaverse platforms, requiring governance that ensures responsible and ethical decision-making from the outset. Deployment concerns the operation and ongoing management of these platforms, which also involves ethical challenges. Although development and deployment often overlap, it is helpful to treat them separately. In development, primary responsibility lies with platform developers, while secondary actors (e.g., regulators and standards bodies) provide oversight and influence. In deployment, platform operators are primarily responsible for responsible platform management, while other stakeholders play supporting roles. These distinctions are crucial in assigning responsibilities across the metaverse's life cycle.

6.2.2 Corporate-Driven Responsible Governance

If the metaverse follows the trajectory of the internet, its growth will largely be led by private companies, while national governments and international organizations provide overarching frameworks. This dynamic results in a governance model where corporations bear primary responsibility for both development and operation. We refer to this as *corporate-driven responsible governance*. This model relies on corporate social responsibility (CSR), a strategic business approach expressing a company's commitment to ethical operations and consideration of stakeholder interests beyond profit maximization. CSR is often seen as the operational arm of corporate ethics, reflecting the moral principles that shape business conduct.

CSR has gained wide adoption since the 2000s. According to Matten (2012), companies engage in CSR for four key reasons:

1. *Profitability:* CSR can improve stakeholder relations, build public trust, and prevent restrictive regulations. It also enhances corporate appeal to customers, employees, and investors—particularly in the age of ESG (Environmental, Social, and Governance) investing, where ethical performance influences capital flow.
2. *Stakeholder Management:* Businesses face competing demands from multiple stakeholders. CSR provides a strategic framework for managing these relationships and making decisions that balance diverse interests.
3. *Moral Navigation:* CSR helps companies address societal concerns and resolve internal ethical dilemmas. It provides a foundation for decision-making aligned with societal values.
4. *Corporate Citizenship:* As companies grow in power—sometimes rivaling governments—CSR provides a way for them to embrace a civic role, acknowledging responsibilities to the broader global community.

The case for CSR does not guarantee that companies will act responsibly. Profit motives—short- or long-term—can conflict with responsible practices, which may raise costs, limit market opportunities, or slow innovation. For instance, investments in fair labor, sustainable technologies, or compliance measures can reduce competitiveness, especially when competitors do not make similar commitments. These tensions can discourage sustained CSR investment, especially when economic conditions are unfavorable. For example, during the 2010s and early 2020s, many artificial intelligence (AI) companies established ethics teams to manage risks. However, by 2022–2023, economic pressures led to widespread layoffs in these teams, even as generative AI technologies raised new ethical challenges. One major tech firm reportedly reduced its ethics team to accelerate product releases in a competitive landscape—highlighting the fragility of CSR when profit dominates.

6.2.3 The Role of Other Stakeholders

Given the limitations of relying solely on corporate responsibility, other actors must contribute to responsible metaverse governance. These stakeholders can either support responsible corporate practices or take independent action to ensure ethical outcomes.

6.2.3.1 Governments

Because profit-driven companies may scale back responsible practices when they threaten competitiveness or profitability, voluntary CSR alone is often insufficient to ensure socially beneficial outcomes. This creates a need for government

6.3 The Role of Other Stakeholders

intervention. Through regulation, governments can establish baseline standards that all companies must meet, reducing the incentive to cut corners and leveling the playing field. Regulatory measures help ensure that fundamental concerns—such as individual rights, environmental sustainability, and mental health safeguards—are consistently upheld, even when market pressures or economic downturns might otherwise discourage responsible action.

Governments are tasked with promoting societal well-being and upholding public values. They influence private sector behavior through laws, regulations, and standards addressing areas such as data privacy, labor practices, environmental protection, and stakeholder participation. They can also mandate CSR disclosures and enforce ethical conduct through procurement policies or tax incentives. Furthermore, governments support international organizations (e.g., UNESCO, WTO, ITU) that help define global norms for responsible practice.

6.2.3.2 Trade Associations

Trade associations represent businesses within a specific industry. They can promote responsible practices by developing shared codes of conduct, offering training and support, and coordinating CSR efforts across companies. They also help address the competitive disadvantage that can deter individual companies from adopting CSR, creating an industry-wide level playing field. However, these associations ultimately serve member interests and may resist measures perceived as burdensome.

6.2.3.3 Standards Organizations

These bodies establish technical and ethical standards for products, services, and business practices. Some also certify compliance. Key organizations include the International Organization for Standardization (ISO), the Internet Engineering Task Force (IETF), and the International Electrotechnical Commission (IEC). By promoting interoperability, transparency, and accountability, their standards influence both corporate behavior and regulatory frameworks.

6.2.3.4 Universities

Universities shape future professionals and play a critical role in promoting ethics through education and research. This includes teaching business ethics, technology ethics, and research integrity. Universities are also central to responsible innovation, often supported by funding bodies that encourage ethical review, inclusivity, and public engagement.

6.2.3.5 Civil Society and Media

Civil society organizations—ranging from consumer advocacy groups to privacy watchdogs and human rights NGOs—help hold corporations and governments accountable. Their influence comes through advocacy, research, partnerships, and public campaigns. They often draw attention to ethical concerns and push for stronger protections. Meanwhile, media organizations play a watchdog role, raising public awareness, exposing misconduct, and highlighting positive examples of responsible practice. Together, these stakeholders form a complex ecosystem that can support responsible ethical practices in metaverse development and use.

6.2.3.6 Alternative Models of Responsible Governance

Many metaverse advocates and stakeholders envision a future metaverse that is open, decentralized, and not dominated by for-profit corporations. They argue that when corporate interests take precedence, concerns about social responsibility and public welfare are often sidelined in favor of market competitiveness. In response, they promote alternative governance models in which non-corporate actors—such as nonprofits, cooperatives, and community-led initiatives—take the lead in developing, owning, and managing metaverse platforms. In the previous chapter, we explored several such models with an emphasis on democratic governance. Here, we return to these models with a particular focus on how they can enable more responsible and socially accountable governance.

One such model is decentralized governance, where decision-making is distributed among users who propose and vote on policies. They are also individual or collective owners of (parts of) the metaverse. This model minimizes tensions between profit and public interest, as community values may directly shape policies. It could lead to stronger protections for rights such as privacy, freedom, and digital property, as well as greater concern for individual and collective well-being.

However, this model is not without risks. If user decisions are driven by narrow self-interest, policies may lack broader ethical consideration. A "tyranny of the majority" could emerge, marginalizing minority or niche groups. Non-users—who are unrepresented—may be negatively affected by policies focused solely on users. Moreover, if participation is limited by technical, economic, or social barriers, governance may fall into the hands of a vocal or privileged minority. To mitigate these risks, decentralized governance must be deliberative and grounded in shared principles. Broader participation should be enabled, and non-user interests should be represented to ensure alignment with societal well-being.

In Chap. 5, we also discussed centralized models involving representative democratic governance under nonprofit, corporate, or hybrid ownership. In nonprofit-led models, the absence of profit pressures supports more responsible governance. In corporate or mixed models, tensions may persist but are moderated by participatory structures aligned with stakeholder values. These centralized models face similar risks as decentralized ones—majority dominance, exclusion of non-users, and limited participation. As such, they require similar safeguards to ensure inclusive, fair, and responsible governance of the metaverse.

6.3 Responsible Development and Ethics by Design

In the previous section, we distinguished between two main governance processes for the metaverse: for development and for deployment and operation. This section focuses on responsible development and the strategies and tools that can support it. A key assumption we will make is that technology is not morally neutral. Design choices carry significant social consequences, and responsibility must extend to the development phase—not just to later use. While widely accepted in technology ethics, this view still warrants clarification.

The claim that technology is neutral is often supported by the idea that tools can be used for both good and bad purposes. A hammer can build a house or cause harm. Social media can foster connection or spread misinformation. In this view, responsibility lies solely with users. However, modern technologies are not simple tools. They embed specific functions and effects that users cannot easily override. A gasoline car inevitably emits greenhouse gases. Most social media platforms inherently collect personal data due to their design. An AI system trained on biased data will likely produce biased outcomes, regardless of user intent. These examples show that moral impacts are built into many technologies and emerge from design itself—not just from use.

This has clear implications for governance. Technology development processes must aim to embed appropriate values into technological design. Since the 1980s, a range of design methodologies has emerged to meet this goal, including value-sensitive design (Friedman et al. 2006), Design for Values (van den Hoven et al. 2015), and Ethics by Design (Brey and Dainow 2023). Additional approaches include Privacy by Design, Secure by Design, and algorithmic fairness.

Responsible metaverse development can also be supported by ethics guidelines, codes of conduct, assessment tools, co-creation, standards, regulation, and CSR strategies. Yet among these, ethical design approaches are the most essential, as they directly integrate moral concerns into design decisions. Other tools, while helpful, often lack the specificity needed to guide developers in translating abstract values—like freedom or fairness—into concrete design choices.

6.3.1 Ethics by Design

A central instrument for the responsible development of the metaverse is the approach known as *Ethics by Design*. Ethics by Design, as developed in Brey and Dainow (2023), is built on the idea of embedding moral values and solutions in designs during the development process. Its premise is that developers cannot be expected to improve the ethical quality of a system through abstract principles or general assessments alone. They require actionable tools that embed ethical considerations directly into the design methodology. Ethics by Design provides such tools by specifying concrete tasks to be carried out during each stage of system development. These tasks address common, recurring ethical issues; more complex concerns may still require the involvement of an ethicist, who can support development teams through initial risk assessments and structured deliberation.

Ethics by Design is based on the observation that the development of information systems typically unfolds in six phases, whether through sequential (e.g., waterfall) or iterative (e.g., agile) processes. These phases are:

1. specification of objectives
2. specification of requirements
3. high-level design
4. data collection and preparation
5. detailed design
6. testing and evaluation

In Ethics by Design, each of these phases is associated with specific tasks designed to align development activities with ethical goals, such as fairness, privacy, accountability, and transparency. Some tasks are generic—such as stakeholder involvement or ethical risk assessment—while others target specific principles, such as minimizing algorithmic bias or enabling auditability.

Ethics by Design relies in part on existing ethics guidelines but translates them into *ethics requirements*: high-level system specifications that are aligned with ethical principles. These requirements function analogously to technical design requirements and provide a bridge between abstract values and implementable design actions. For example, a general principle such as fairness may be expressed through more specific requirements, including universal accessibility or the avoidance of algorithmic discrimination.

Typical responsibilities at each phase can be summarized as follows:

1. *Specification of objectives:* At this early stage, system goals are defined. Ethics-related tasks include reviewing whether these objectives align with ethics requirements and consulting stakeholders to ensure that broader social values are taken into account.
2. *Specification of requirements:* This phase defines technical and non-technical requirements and identifies needed resources. Ethics-related tasks here include integrating ethics requirements into the overall requirements set, evaluating the compatibility of other requirements with ethical goals, soliciting stakeholder input, conducting ethical risk and impact assessments, and drafting an Ethics by Design implementation plan.
3. *High-level design:* In this stage, the system's architecture is developed. Tasks include ensuring that design choices support ethics requirements or, where necessary, modifying them to resolve potential conflicts.

4. *Data collection and preparation:* This involves gathering and processing the data that will be used in the system. Ethics-related tasks focus on ensuring that datasets are free from harmful bias, respect privacy, and support accountability.
5. *Detailed design:* As the system is implemented, features must be added or tested to ensure they support the established ethics requirements. This may include interface elements, access controls, and other system functionalities.
6. *Testing and evaluation:* In the final phase, the system is assessed for performance, including ethical performance. Tasks include verifying compliance with ethics requirements, gathering stakeholder feedback on whether the system reflects their values, and, where relevant, reviewing supporting materials such as documentation, complaints procedures, and user training.

6.3.2 Applying Ethics by Design to the Metaverse

Application to the metaverse involves three primary steps. First, general ethical guidelines for the metaverse must be established. See the annex of this book for a proposed set of such guidelines. Second, these guidelines must be translated into concrete ethics requirements tailored to the design and operation of metaverse platforms. Third, specific ethics-related tasks must be developed for each design phase.

The translation of guidelines into requirements is not mechanical. It involves critical interpretation and depends on a clear understanding of how ethical principles relate to technological design. This process is best undertaken collaboratively between ethicists and developers. For example, a principle that aims to protect freedom rights can be connected to multiple design concerns. In Chap. 4, we explored how both the operation and design of metaverse platforms can affect user freedom. From this, one might derive an ethics requirement to avoid manipulative or deceptive design features and to actively implement mechanisms that counter them.

Such a requirement—e.g., avoidance of deception and manipulation—can be translated into tasks at several design stages. To avoid "dark patterns," for example, developers may need to:

- confirm in the requirements phase that no manipulative features are included,
- ensure during high-level design that no architectural elements necessitate deceptive practices, and
- test during detailed design that user interface elements are free from misleading cues.

In this way, ethical principles are systematically translated into actionable design tasks, ensuring that core values are embedded throughout the development process and not left to interpretation or chance.[1]

---

[1] To observe a detailed version of Ethics by Design, with a full specification of requirements and tasks, see the version we developed for the European Commission for AI systems (European Commission 2021).

6.3.3 The Broader Ecosystem

While developers and engineers play a central role in applying Ethics by Design, successful implementation also depends on organizational and regulatory support. Companies must commit at the management level and integrate Ethics by Design into their broader CSR strategies, internal ethics codes, and existing oversight procedures, such as risk assessments or ethical audits. Adequate resources must be allocated for training and implementation, and Ethics by Design should be embedded in both workflow and culture.

Other actors in the broader governance ecosystem also contribute. Governments can intervene when public interest is at stake, mandating or prohibiting certain design features. For example, they may require universal accessibility, as discussed in Chap. 8, or restrict the use of manipulative interface elements. Standards organizations can likewise promote responsible development by issuing design-oriented ethics standards. The ISO, for instance, has published multiple standards related to AI ethics. In 2023, the IEEE introduced two initiatives specific to the metaverse: the IEEE P7030 standard for ethical assessment of extended reality technologies, and the IEEE P7016 standard for ethically aligned design and operation of metaverse systems.

These regulations and standards, once developed, can be translated into ethics requirements and incorporated directly into the Ethics by Design process. In this way, Ethics by Design serves as both a practical methodology and a governance mechanism that helps ensure that metaverse technologies are aligned with societal values from the ground up.

6.4 Responsible Platform Operation

We now turn to tools and strategies for responsible operation of metaverse platforms. Platform operation encompasses a wide range of tasks, including maintenance, user support, content management, performance optimization, privacy and security management, community building, enforcement of Terms of Service (ToS) and community guidelines, marketing, and financial management. Each of these functions involves ethical decisions that can significantly impact users and stakeholders.

6.4.1 Ethics in Deployment and Operation

To guide responsible operation, we propose an approach called *Ethics in Deployment and Operation* (EDO), which mirrors the structure of Ethics by Design but applies to post-deployment phases. Drawing on leading IT management frameworks such as ITIL (Information Technology Infrastructure Library) (AXELOS 2019) and COBIT (Control Objectives for Information and Related Technologies) (ISACA 2019a; ISACA 2019b), we distinguish five key processes in the lifecycle of IT systems:

1. IT Management Strategy
2. Acquisition and Design
3. Deployment and Implementation
4. Service Operation
5. Monitoring, Assessment, and Improvement

Among these, service operation is most critical, as it is the phase during which users interact with the system and most societal impacts occur. However, all phases involve decisions that influence ethical outcomes during operation. For instance, a failure to include strong privacy features in the acquisition phase can compromise user data during operation.

EDO assigns specific ethics tasks to each of the five IT management processes. These tasks are designed to meet ethics requirements derived from broader ethical guidelines. Below is a brief overview of ethics tasks for each phase in the IT system lifecycle. Their application to metaverse platforms will be addressed thereafter.

1. *IT Management Strategy*
   This process involves defining IT governance objectives, policies, and organizational structures. Ethics tasks include establishing a formal ethics strategy aligned with CSR, assigning staff roles for implementation, and developing policies for monitoring and compliance. Ethics training and a culture of responsibility should also be promoted.
2. *Acquisition and Design*
   This phase includes assessing whether proposed solutions meet ethical guidelines. If building in-house, an Ethics by Design approach is recommended. If outsourcing, systems should be evaluated for ethical compliance. Risk and impact assessments should be completed before proceeding.
3. *Deployment and Implementation*
   This phase introduces the system into its operational environment. Ethics-related tasks include integrating the ethics strategy into usage policies, updating security and access protocols to align with ethics requirements, embedding ethical content in training, and engaging stakeholders in implementation efforts.
4. *Service Operation*
   Here, the platform is fully functional. Ethical responsibilities include enforcing ethics policies among staff and users, monitoring emerging ethical issues, and ensuring equitable access and responsible content moderation.
5. *Monitoring, Assessment, and Improvement*
   Continuous evaluation ensures ethical performance and compliance. Ethics-related tasks include adding ethics metrics to monitoring frameworks, conducting audits, and engaging qualified assurance providers with expertise in ethical compliance.

6.4.2 Applying EDO to Metaverse Platforms

To apply EDO to the metaverse, three steps are required: (1) establishing general ethics guidelines, (2) translating them into ethics requirements for platform operation, and (3) specifying ethics-related tasks for each IT lifecycle phase, adapted to the context of metaverse operations. Metaverse platforms are not merely support systems; they are core to a company's business model, delivering immersive environments and social interactions. As such, ethical considerations should be integrated into both IT and overall business strategies. Business plans must be assessed against ethics guidelines from the outset.

In the annex, we propose ethics guidelines for the metaverse. They include nine main principles. Below, we highlight how each of them can be operationalized in platform operation.

- **Security**
  Security encompasses both cybersecurity and virtual security. As discussed in Chap. 5, preventing harmful or criminal behavior in virtual environments requires proactive design and enforcement. This includes robust ToS, technical safeguards, and active moderation.

- **Freedom**
  Freedom-related principles should be addressed through user rights protections, moderation practices, and policies that limit manipulative design and misinformation. ToS must include protections for autonomy and support mechanisms for redress.

- **Privacy**
  Privacy should extend beyond data protection to include limits on monitoring user behavior, avatars, and digital assets. This requires technical safeguards and clear policies restricting surveillance by both platform operators and other users.

- **Equality, Fairness, and Inclusion**
  These values can be advanced by fostering a diverse and inclusive company culture and ensuring universal access through assistive technologies and user support. Platforms should assess for bias in content, algorithms, and policies, and take action to mitigate exclusionary dynamics. Training programs and inclusive community events can further support these goals.

- **Property**
  Operators should protect users' digital, virtual, and intellectual property via clear provisions in the ToS and End-User License Agreements (EULA). Ethical property management includes ensuring fair economic interactions, securing virtual assets, and preventing market manipulation. In some cases, collaboration with law enforcement may be required to combat fraud or theft.

- **Well-being**
  Supporting user well-being includes minimizing harmful or addictive platform design, providing education on healthy use, and offering tools like parental controls. Ideally, platforms should include staff dedicated to digital well-being and regularly monitor user experiences through feedback or surveys.

- **The Common Good**
  Platforms should foster community development through collaborative tools, public events, and integration with real-world social connections. Operators should also mitigate threats to community integrity, such as toxic behavior, misinformation, or harmful ideologies. Close collaboration with user communities and civil society groups can support this work.

- **Environmental Sustainability**
  Platform operators can promote sustainability by using environmentally responsible cloud services, infrastructure, and supply chains. They can influence user behavior through awareness campaigns and promote virtual alternatives to resource-intensive physical activities. Digital twins can also be used to support environmental monitoring and design.

- **Accountability and Governance**
  Operators must embed accountability into their organizational structure and work culture. This includes clear governance bodies, transparency mechanisms, regular audits, and stakeholder engagement processes. Governance should include user and stakeholder representation, particularly in decisions affecting virtual environments and user experience. A strong CSR strategy, supported by ongoing training and inclusive decision-making, can help align operations with broader societal values.

6.4.3 The Role of Other Actors

While platform operators bear primary responsibility for ethical operation, other actors also play essential roles. *Governments* can regulate content moderation, security, privacy, and inclusion. Legislation may define standards for user protection, especially for vulnerable populations. *Trade associations* can develop industry standards and promote self-regulation, providing frameworks that complement public regulation and build sector-wide consensus. *International organizations* such as the Internet Governance Forum, World Trade Organization, and International Telecommunication Union can advance non-binding standards or policy coordination that influence platform practices globally. Finally, *users and user communities* are crucial for responsible governance. They can contribute through feedback, advocacy, and active participation in governance forums. Their insights help ensure that policies reflect lived experiences and evolving expectations.

6.5 Conclusion

In this final chapter, we examined how ethical issues in the metaverse can be addressed through responsible governance, emphasizing the roles and responsibilities of various actors. Section 6.2 introduced the need for multistakeholder governance to achieve ethical outcomes, followed by a discussion of governance principles and models, including both corporate-led and alternative frameworks, with attention to corporate social responsibility. Section 6.3 focused on responsible development, highlighting *Ethics by Design* as a structured approach for embedding ethical values in design processes. Section 6.4 extended this framework to the operational phase, proposing the *Ethics in Deployment and Operation* (EDO) model and applying it to metaverse platforms.

Ethics guidelines are an important tool for responsible practice and stakeholder coordination. In the annex, a proposal is made such for guidelines.

As emphasized in this book, the metaverse has the power to reshape society and daily existence. The social ethical implications associated with the metaverse are substantial, necessitating involved stakeholders to make informed choices, and promote responsible practices throughout the entire process of development, operation, and governance. Within these pages, we have delved into the most critical ethical and societal concerns presented by the metaverse and have propose strategies for effectively mitigating them. We hope that this book will foster critical debate and encourage the adoption of responsible practices in the development, operation, and governance of the metaverse—whatever form it may take and whatever timeline it may follow.

References

AXELOS. 2019. *ITIL® Foundation, ITIL*. 4th ed. TSO (The Stationery Office).

Brey, P., and B. Dainow. 2023. Ethics by Design for Artificial Intelligence. *AI Ethics* 4:1265–1277. https://doi.org/10.1007/s43681-023-00330-4.

European Commission. 2021. *Ethics by Design and Ethics of Use Approaches for Artificial Intelligence (Version 1.0)*. European Commission DG Research and Innovation, November 25. https://ec.europa.eu/info/funding-tenders/opportunities/docs/2021-2027/horizon/guidance/ethics-by-design-and-ethics-of-use-approaches-for-artificial-intelligence_he_en.pdf. Accessed 1 April 2025.

Friedman, Batya, Peter H. Kahn Jr., and Alina Huldtgren. 2006. Value Sensitive Design and Information Systems. In *Human–Computer Interaction in Management Information Systems: Foundations*, ed. Ping Zhang and Dennis Galletta, 348–372. Armonk, NY: M.E. Sharpe.

ISACA. 2019a. *COBIT 2019 Framework: Introduction and Methodology*. Schaumburg, IL: ISACA.

ISACA. 2019b. *COBIT 2019 Framework: Governance and Management Objectives*. Schaumburg, IL: ISACA.

Matten, Dirk. 2012. Why Do Companies Engage in Corporate Social Responsibility? Background, Reasons and Basic Concepts. In *The ICCA Handbook on Corporate Social Responsibility*, ed. Judith Hennigfeld, Manfred Pohl, and Nick Tolhurst, 1–46. Chichester: Wiley.

Van den Hoven, Jeroen, Pieter E. Vermaas, and Ibo van de Poel. 2015. *Handbook of Ethics, Values, and Technological Design*. Dordrecht: Springer.

Annex: A Moral Framework for Metaverse Ethics

When emerging technologies raise significant social and ethical concerns, various organizations often respond by issuing ethical guidelines. In the case of AI, for example, such guidelines have been developed by companies, national governments, the IEEE, the EU, OECD, and UNESCO. These documents aim to promote responsible behavior and provide a shared foundation for stakeholder dialogue and decision-making.

However, ethics guidelines have limitations. They are typically broad and open to interpretation, often lacking mechanisms for implementation and enforcement. They may offer little help in addressing value conflicts or complex dilemmas and can be used for "ethics washing"—superficial commitments that improve image without changing practice. Ethics guidelines should be seen as one among several tools for responsible technology governance. They are valuable when paired with efforts to operationalize and enforce them, but they cannot be mechanically applied to resolve ethical issues. Identifying such issues requires moral sensitivity, and resolving them requires critical reflection and deliberation.

In this annex, a proposal is made for general ethics guidelines for the metaverse, directed not at any single actor but at society as a whole. They apply to all aspects of the metaverse—its development, deployment, use, and governance. They consist of nine overarching principles, each supported by subprinciples offering more specific direction. They are grounded in the previous discussion of rights, well-being and the social good in Chaps. 4 and 5 of this book, as well as international human rights norms and widely accepted ethical standards.

Ethics Guidelines for the Metaverse

The metaverse should be developed, deployed, and operated with the aims of providing benefits to individuals and society and upholding human values and standards of ethics. In particular, the following principles should be adhered to:

1. *Safety and security*

    In the metaverse, safety and security should be paramount. The right to security of persons should be guaranteed, and metaverse assets and infrastructure should be protected from unauthorized access, theft, tampering, and loss.

    a. Metaverse technology should be safe to use and free from risks to physical health, mental well-being, and cognitive development.
    b. Metaverse systems must uphold strong cybersecurity standards to ensure the security and integrity of data and system resources.
    c. Measures should protect the metaverse from national security threats and safeguard digital twins and IoT-integrated platforms from cyber-physical attacks.
    d. Users must be protected from crimes against persons, including harassment, assault, bullying, identity theft, sex crimes, and exploitation.
    e. Protections should be in place for digital and virtual property, guarding against theft, fraud, vandalism, market manipulation, and other property-related crimes.
    f. Systems must also address broader crimes, such as the spread of illegal content, unauthorized gambling, illicit trade, and criminal coordination.
    g. Children should only be allowed to access the metaverse when proven safe for their health and cognitive development.
    h. Children require enhanced protections due to their vulnerability to bullying, abuse, exploitation, inappropriate content, and recruitment into virtual crime.

2. *Freedom*

    Freedom rights should be protected in the metaverse, including freedom of expression, peaceful assembly and association, freedom of movement and residence, freedom of thought and belief, bodily integrity, and autonomy.

    a. Freedom of expression should be upheld, encompassing freedom of information and freedom of speech, which includes artistic expression and immersive simulations used for information, inspiration, and entertainment. Restrictions may apply to speech that risks imminent harm, violates rights, or is offensive without redeeming value.
    b. Anonymous speech should be allowed where possible but may be limited for security reasons.
    c. The right to bodily integrity, which is the right to control one's body and be free from unwanted intrusion or contact, extends to avatars when users identify with them and experience a strong sense of embodiment and control.

Annex: A Moral Framework for Metaverse Ethics

   d. Freedom of movement and residence should be upheld in the metaverse, including rights to move within and between virtual worlds and to access, own, or rent virtual property.
   e. Peaceful assembly and association must be respected and should not be restricted without compelling reason and due process.
   f. Autonomy and self-determination should be protected by shielding users from manipulation and deception, including disinformation, social engineering, false identities adopted by users and chatbots, dark patterns, and covertly personalized experiences.

3. *Privacy*

   In the metaverse, privacy should be protected, including the privacy of users and data subjects whose personal data are stored and processed on metaverse platforms.
   a. Privacy in the metaverse should cover not only information privacy but also bodily, spatial, proprietary, intellectual, decisional, associational, and behavioral privacy. It should protect against virtual interference, intrusive observation, and excessive data collection or use.
   b. Platforms should adopt a privacy by design approach, embedding privacy and data protection as core features.
   c. Decentralized data storage models using blockchain and self-sovereign identity should be actively explored.
   d. Personal data should follow standard privacy principles, including notice, consent, purpose limitation, data minimization, security, accuracy, limited retention, and accountability.
   e. Informed consent must be meaningful, with clear, accessible explanations of the consequences for the user's interests.
   f. Sensitive data—biometric, health-related, emotional, or ideological—should only be collected with consent for restricted, clearly defined purposes serving user or public interest.
   g. Data on private virtual entities—such as owned buildings, private spaces, conversations, or associations—should only be collected with consent and for limited, justified uses.
   h. Data segregation policies should prevent reuse of data for purposes not originally consented to.
   i. Third-party data sharing requires informed consent. Platforms must ensure contractual and verifiable adherence to proper data use standards.
   j. Personalization should not be based on highly sensitive data, and users should not be confronted by persuasive messaging that references their private affairs.
   k. User privacy must be protected from violations by other users through technical safeguards and appropriate penalties.
   l. Data re-identification should be prevented through strong cybersecurity, limited use of behavioral biometrics, and risk mitigation strategies.

m. Microtargeting, virtual influencers, and product placement must not deceive or manipulate users by using analytics that exploit political and ideological beliefs, personality traits, or emotions to get a response that bypasses rational thought processes.

4. *Equality, Fairness and Inclusion*

In the metaverse, the right to equality should be protected, users and other stakeholders should be treated fairly, and diversity, equity, and inclusion should be prioritized.

   a. The metaverse should support equality by ensuring equal rights and opportunities, equal protection from discrimination, equal access to services, and equal pay for equal work.
   b. It should also promote equity by providing additional resources and opportunities to disadvantaged or special-needs users to support their success.
   c. Diversity and inclusiveness should be fostered by enabling full participation of users with different identities and by valuing differences positively.
   d. The metaverse should be developed and managed by diverse teams within an inclusive workplace culture.
   e. The metaverse should be accessible to all by removing barriers related to physical access, skills, or usage opportunities. This includes universal connectivity, affordable and safe hardware and software, accessibility tools, language support, and assistance for low-education users.
   f. Social biases in design and operations—such as access, skill, functional, algorithmic, and representational biases—should be actively mitigated. This requires bias-aware design, fairness assessments of policies, and diverse, inclusive design and support teams.
   g. User-to-user discrimination—including hate speech, harassment, hate crimes, and exclusionary practices—should be addressed through inclusive community building, training, and enforcement of anti-discrimination policies.
   h. The use of VR and AR applications to foster empathy and reduce prejudice should be considered for education, training, and integration into gaming.

5. *Property*

The right to property—including digital, virtual, and intellectual property—should be protected in the metaverse. Ownership models should uphold this right while also serving the common good.

   a. Metaverse platforms must include clear regulations to ensure safe ownership, use, sale, gifting, renting, collateralization, and inheritance of property.
   b. Rights to virtual property should be explicitly defined and protected.
   c. Intellectual property rights must be upheld, especially for user-generated content and NFTs.
   d. Online markets should be safeguarded from manipulation and deceptive practices, protecting both creators and buyers.

e. Consumer protections should guard against misleading advertising that uses virtual simulations or product placements to misrepresent products and services.
f. Ownership models should promote the common good and be evaluated for their broader impact.
g. The metaverse should enable users to create, acquire, and assert ownership over digital and virtual goods, including intellectual property rights.
h. The metaverse should accommodate private, public, and collective ownership of virtual and digital assets.
i. Use of cryptocurrencies in the metaverse should include safeguards against fraud, theft, and criminal activity, with attention to inequality and environmental impact.
j. Monetization of personal data should only occur where risks to privacy, autonomy, and democracy can be effectively mitigated.

6. *Well-being*

The metaverse should be designed and operated with a strong focus on user well-being, with safeguards to prevent harm.

a. The metaverse should be developed and operated so as to support key aspects of well-being, including pleasure, physical and mental health, social connection, autonomy, achievement, meaningfulness, and personal growth.
b. Design and governance should be guided by research on how metaverse and internet use affect mental health, with attention to risks like depression, anxiety, low self-esteem, body image issues, loneliness, and addiction—especially among youth.
c. To prevent compulsive use and addiction, evidence-based measures should be implemented, particularly for high-risk applications like gaming, gambling, social media, and pornography. Addictive design practices must be avoided.
d. The metaverse should promote social well-being by supporting strong relationships and community. New features should be evaluated for their impact on relationships, and steps should be taken to reduce harms such as displacement, infidelity, deception, and unrealistic expectations about relationships.

7. *Social quality and the common good*

The metaverse should be developed and governed with a focus on promoting and maintaining social quality and social goodness within the metaverse and in society at large.

a. The metaverse should be developed and governed to promote social quality, supporting socio-economic security, social cohesion, inclusion, and empowerment.
b. Impacts on society should be taken into account in developing and governing the metaverse, such as implications for the economy, employment, political processes, education, healthcare, social cohesion, public safety, social trust, and civic engagement.

8. *Sustainability*

    The metaverse should be developed, operated, and used in line with environmental sustainability goals, aiming to reduce its ecological impact and to support sustainable practices in society.

    a. The metaverse should be developed and operated according to principles of sustainability, circular economy and energy efficiency, which should be employed throughout its lifecycle.
    b. Data centers for the metaverse should adopt green energy, improve energy and water efficiency, and apply circular economy practices.
    c. Virtual replacements of environmentally harmful physical activities should be prioritized, alongside measures to limit rebound effects.
    d. Digital twins should be used for eco-friendly design, monitoring, and optimization of human-made systems and natural environments.
    e. Immersive simulations should support environmental education, raise awareness, and build skills for addressing ecological challenges.

9. *Accountability and responsible governance*

    Robust frameworks for responsibility and accountability should be established to foster a culture of responsible development, deployment, and use of the metaverse. These should be embedded within a broader governance framework that promotes ethical practices, social responsibility, democratic participation, and transparency.

    a. Developers should take responsibility for the societal impacts of metaverse technologies and platforms, addressing harm through corporate social responsibility (CSR) and responsible innovation strategies.
    b. Platform operators must be accountable for the consequences of metaverse operations and should invest in CSR and responsible deployment practices.
    c. Governments should enact and enforce regulations and policies that promote responsible development and operation of the metaverse and mitigate harms.
    d. Users are responsible for their behavior in the metaverse. They should respect others' rights, follow community guidelines, avoid harmful activity, promote inclusivity, make positive contributions, and participate in democratic governance.
    e. Other key actors—such as trade associations, civil society groups, standards bodies, universities, and media—should contribute to responsible metaverse governance within their domains.
    f. Transparency should be ensured through accessible disclosures about metaverse operations, including algorithms, data use, policies, security, and governance structures.
    g. Governance of the metaverse should be democratic, allowing meaningful participation for users and stakeholders, either through direct democratic processes, where users take on accountability themselves, or through representative governance, where elected actors are accountable through voting and civic participation.

Commentary to the Guidelines

The first eight principles reflect *substantive values*—values that are intrinsically important, such as freedom and well-being. The ninth principle, *accountability*, is procedural. It serves as a mechanism to uphold the substantive values by ensuring that actors take responsibility for their actions and are answerable, particularly when their actions have moral consequences. Accountability here refers specifically to *public accountability*—being transparent, responsive, and answerable to society. For companies, this includes fulfilling CSR and stakeholder commitments. For public and civil society organizations, it involves responsible stewardship of public interests and resources.

We considered including a tenth principle—*truthfulness*—given that metaverse environments can facilitate deception through impersonation, counterfeiting, deepfakes, and hyper-real simulations. Few technologies enable deception in so many ways. However, truthfulness is not universally applicable in the metaverse. Many virtual environments are intentionally fictional, designed for entertainment, role-play, and artistic expression. In such contexts, strict standards of truth are not expected. Moreover, the essence of VR lies in sensory illusion. For these reasons, truthfulness is not proposed as a separate principle. Nonetheless, concerns about deception are addressed through subprinciples under *freedom* (which includes autonomy and self-determination) and *accountability* (particularly transparency, which allows users to assess the reliability of content and actors).

Index

A
Access, 74–75
Accountability, 121, 128
Addiction, 44, 90
Addiction by design, 90, 91
Algorithmic bias, 76
Anticipatory Technology Ethics (ATE), 4
Artificial intelligence (AI), 2, 25–26
Augmented reality (AR), 16, 23–25, 37
Autonomy, 68
Avatar, 23, 48

B
Being, 121
Benefits, 2, 3, 35–41, 44, 86
Bias, 75–76
Blockchain, 26, 47, 81
Bodily integrity, 67
Brain-computer interfaces (BCIs), 23

C
Cloud computing, 28
Commerce, 39–40
Common good, 80–81, 93, 97, 121, 127
Communities, 94
Content moderation, 65–67
Corporate social responsibility (CSR), 111
Cryptocurrencies, 81–82
Cybersecurity, 61

D
Data protection, 69, 125
Decentraland, 51
Decentralization, 19
Decentralized governance, 95
Dematerialization, 99
Democracy, 95–96
Democratic governance, 95–96
Design, 115–117
Digital economy, 12
Digital identity, 19
Digital property, 61
Digital twins, 21, 27, 102
Discrimination, 76–77
Diversity, 74, 126

E
Edge computing, 11, 28
Education, 38
Embodied, 6, 92
Embodiment, 43, 48, 91
Entertainment, 38
Environmental risks, 98
Equality, 73, 120, 126
Ethics by Design, 115–117
Ethics guidelines, 120
Ethics in Deployment and Operation (EDO), 118–121
Extended reality, 25

F
Fairness, 74, 120
Family, 91–92
5G/6G, 28
Freedom, 65, 120, 124
Freedom of expression, 65
Freedom of movement and residence, 67
Friendship, 92
Functional bias, 76

G
Geographical disembedding, 52–53
Governance, 19, 95, 110–115, 121, 128

H
Harassment, 64, 77
Harm, 60
Hate crimes, 77
Hateful content, 76
Healthcare, 18, 40
Human-computer interaction (HCI), 47–48
Human rights, 60

I
Identity theft, 62
Immersion, 17
Immersiveness, 17
Inclusion, 73, 120
Inclusiveness, 74
Internet, 2, 8, 11, 19, 27, 46–47, 88, 90–91, 100
Internet of Things, 27, 47
Interoperability, 15
IT infrastructure, 28

J
Justice, 74

M
Manufacturing, 40
Massively multiplayer online games (MMOs), 49–50
Media, 38
Mental health, 88
Meta, 6–7, 12
Metaverse, 28
  advanced metaverse, 34
  audiovisual metaverse, 17
  gaming metaverses, 18
  hyperreal/hyper-realistic metaverse, 17, 35
  industrial metaverse, 18
  non-immersive virtual worlds and environments, 17
  pan-sensory metaverse, 17
Microsoft, 7, 11, 12
Mixed reality (MR), 25
MMO, *see* Massively multiplayer online games (MMOs)
MR, *see* Mixed reality (MR)
Multifunctionality, 18
Multi-User Dungeons (MUDs), 49, 52

N
Natural interaction, 47–48
Non-fungible tokens, 82
NVIDIA, 7, 11, 12

O
Office work, 40
Openness, 18

P
Peaceful assembly and association, 67
Persistence, 15
Platform governance, 81
Platform operation, 118–121
Privacy, 68, 120, 125–126
Property, 78, 120, 126
Property ownership, 80

R
Rebound effects, 101–102
Regulation, 72, 81–83, 113
Representational bias, 76
Responsible governance, 4, 110
Rights, 3
Risks, 2, 60
  access, 74–75
  bias, 75–76
  censorship, 66
  crimes against property, 64
  crimes against society, 64
  cybercrimes (*see* Cybersecurity threats)
  cyber-physical threats, 62
  dark patterns, 68
  data theft, 62
  deception and manipulation, 68

discrimination, 76–77
disinformation, 68
environmental risks, 98
fake identities, 68
human trafficking, 63
identity theft, 62
metaverse addiction, 89
personalization, 71
personalized reality, 68
sex offenses, 63
social engineering, 62, 68
surveillance capitalism, 71–72
threats to virtual security, 62 (*see* Virtual crimes)
user profiling, 71
user tracking, 71
virtual rape (*see* Sex offenses)
Roblox, 50
Romantic relationship, 92

S
Safety, 44, 124
Second Life, 10, 18, 50
Security, 60, 120, 124
Semantic Web, 46
Sex, 63, 92
Sexual harassment, 63
Social quality, 93, 127
Stakeholders, 112
Standards, 12–14, 18–19, 113, 118
Stephenson, N., 7

Surveillance, 70–72
Surveillance capitalism, 71–72
Sustainability, 102–103, 121, 128

T
3D modeling and animation, 21
Transportation, 41, 100

U
User acceptance, 44
User-generated content, 10
User profiling, 71
User tracking, 71

V
Virtual economy, 51
Virtual property, 79–81, 126
Virtual reality, 7, 21–23
Virtual security, 61
Virtual world, 14, 48

W
Web3/Web 3.0, 19, 46–47
Well-being, 86–88, 127

Z
Zuckerberg, 6, 10–13

GPSR Compliance

The European Union's (EU) General Product Safety Regulation (GPSR) is a set of rules that requires consumer products to be safe and our obligations to ensure this.

If you have any concerns about our products, you can contact us on

ProductSafety@springernature.com

In case Publisher is established outside the EU, the EU authorized representative is:

Springer Nature Customer Service Center GmbH
Europaplatz 3
69115 Heidelberg, Germany